FOOD SCIENCE AND TECHNOLOGY

LEGUMES FOR GLOBAL FOOD SECURITY

FOOD SCIENCE AND TECHNOLOGY

Additional books in this series can be found on Nova's website under the Series tab.

Additional e-books in this series can be found on Nova's website under the eBooks tab.

Food Science and Technology

Legumes for Global Food Security

Alfonso Clemente
and
Jose C. Jimenez-Lopez
Editors

nova science publishers
New York

Library of Congress Cataloging-in-Publication Data
ISBN: 978-1-53612-265-7
Library of Congress Control Number: 2017945941

Published by Nova Science Publishers, Inc. † New York

CONTENTS

PREFACE

Legumes are grown worldwide and comprise the third largest family of flowering plants. Legume species belong to the *Fabaceae* family and are characterized by their fruit frequently named pod. Several of these species have been domesticated such as soybean, common bean, faba bean, pea, lentil, chickpea, lupin, peanut and cowpea. They are a main source of nitrogen-rich edible seeds and constitute a major source of food for a significant part of the world's population, in addition to provide fodder, oil, fiber and many other products.

Globally, grain legumes are one of the most relevant sources of plant proteins, providing also carbohydrates, minerals, vitamins and other components with nutraceutical value and health benefits properties. They also are a valuable crop possessing the ability to fix atmospheric nitrogen by *Rhizobium* biosynthesis and thus contributing to the natural nitrogen cycle, and improving agriculture soil by. The legumes have a vital role to play in both the developed and developing worlds by providing an alternative nitrogen source to nitrogenous fertilizers, which often have had adverse environmental effects.

This book is conceived to provide key research knowledge on the impacts of the N in the environment as consequence of excessive soil fertilization, and overall its influence in the climate change, biocontrol agents to fight phytopathogenic fungi, legume crops productivity, health promoting aspects of seed components, and their nutritional and nutraceutical values, and legume allergy features.

This book presents a series of peer reviewed chapters that the guest editors believe will aid in increasing the quality of the research focus across the growing field of legume research. Overall, the objective of this project is to serve as a reference book and as an excellent resource for students, researchers, and scientists interested and working in different aspects of grain legumes, and particularly for the scientific community to encourage it to continue publishing their research findings on legumes and to provide basis for new research, and the area of sustainable crop production.

We would like to deeply thank the contributors to this project. We are also grateful to the editorial team in Nova Science Publishing for very good cooperation and constant help.

In: Legumes for Global Food Security
Editors: Alfonso Clemente et al.
ISBN: 978-1-53612-265-7

Chapter 1

GRAIN LEGUMES IN SPAIN: THE CURRENT SITUATION AND PERSPECTIVES

Lucía De la Rosa*[1]*, Maria Teresa Marcos*[1]*,
Teresa Millán*[2]*, Josefa Rubio*[3]*, Ana María Torres*[3]*,
***Raquel Olías*[4] *and Alfonso Clemente*[4,*]**
[1]Centro Nacional de Recursos Fitogenéticos, INIA, Madrid, Spain
[2]Universidad de Córdoba, Córdoba, Spain
[3]IFAPA, Centro Alameda del Obispo, Córdoba, Spain
[4]Estación Experimental del Zaidín, CSIC, Granada, Spain

ABSTRACT

Leguminous plants constitute one of the most important botanical families (*Papilionaceae*) from the socioeconomic point of view, playing an essential role in a sustainable agriculture, environmental protection, food safety and nutrition. In this chapter, we report an overview about the current situation and perspectives of grain legumes in Spain, the important role played by a number of Spanish institutions to improve and maintain different collections of genetic resources, the evolution in last

* Corresponding Author Email: alfonso.clemente@eez.csic.es.

decades of yield and cultivated area of the main grain legumes in Spain as well as the importance of the Common Agricultural Policy (CAP) and their effects in the production of legumes in Spain. Finally, data on consumption in Spanish households is reported, where a decrease of 60% have been observed in last decades. This trend must be changed by policies promoting a healthier diet as well as by adaptation to an emerging market oriented towards the development of new products and consumer preferences.

Keywords: CAP, consumers, genetic resources, grain legumes, human nutrition, legume consumption, pulses, Spain

1. Introduction

The 68th Unites Nations (UN) General Assembly declared 2016 the International Year of Pulses (IYP) (http://www.fao.org/pulses-2016). Pulses, also named grain legumes, are those species belonging to the botanical family *Leguminosae* or *Fabaceae* harvested exclusively for dry grain. Some of them are essential for human nutrition as source of proteins and carbohydrates, taking part of a healthy diet. Grain legumes are also useful for feed, as part of the plant-based protein. In this chapter, we will focus the attention on i) the evolution of cultivated area and production of grain legumes in Spain; ii) the genetic resources conservation status of some legume species in Spain used for food and feed; iii) the impact of the Common Agricultural Policy (CAP) on legume crop cultivation; and iv) habits and consumption of legumes in Spain.

2. Evolution of Cultivated Area and Production of Grain Legumes in Spain

Spain, located in the Mediterranean basin, has a long tradition in cultivating different legumes such as pea (*Pisum sativum* L.), chickpea (*Cicer arietinum* L.), common bean (*Phaseolus vulgaris* L.), faba bean

(*Vicia faba* L.), lentil (*Lens culinaris* Medik.), vetch (*Vicia sativa* L.) and bitter vetch (*Vicia ervilia* (L.) Wild.), among others. Unfortunately, over the last decades a severe decrease in terms of cultivated area and production has been observed (Sources: FAOSTAT, 2015; MAPAMA 2016). Changes in cultivated area are the most noticeable with a reduction to half of the area (805.020 ha in 1961 to 437.300 ha in 2016) (Figure 1).

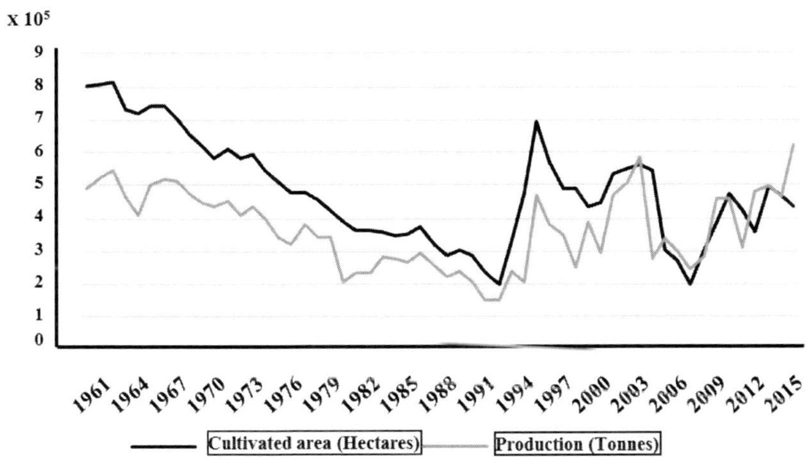

Figure 1. Evolution of grain legumes area (ha) and production (t) in Spain over the last decades (FAOSTAT 2015; MAPAMA 2016).

Several reasons could explain this significant drop, including biotic and abiotic stresses but also the changes in the Common Agricultural Policy (CAP) (see below). The low prices of soybean market and social aspects such as the migration of the population from rural to urban areas could have shaped the current situation. Nowadays, Spain shows a deficit in the market balance of these crops, being currently one of the major importing countries. Grain legumes importations for food have shown a regular increase since the 1980s. In 2015, around 200.000 tonnes of grain legumes were imported, mainly lentils (32%), beans (25,4%) and chickpeas (24,9%) (Figure 2).

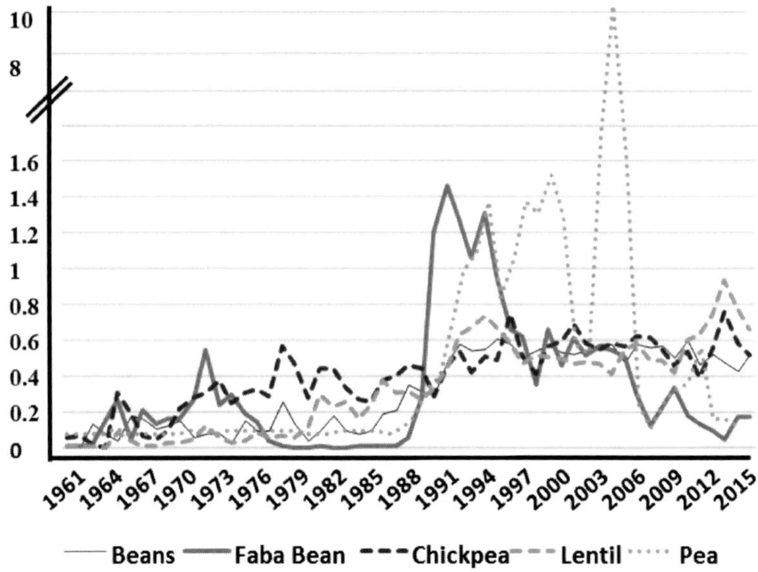

Figure 2. Evolution of main grain legumes imports (t) in Spain (FAOSTAT 2015; MAPAMA 2016).

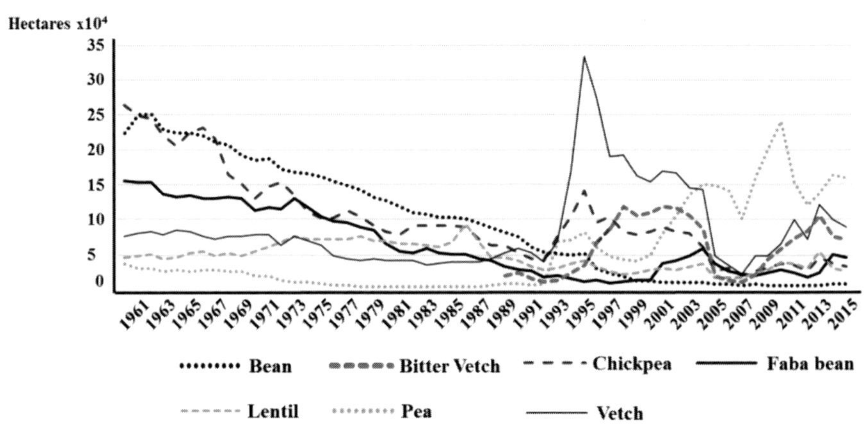

Figure 3. Evolution of harvested area (ha) of main grain legumes cultivated in Spain over the last decades. (FAOSTAT 2015; MAPAMA 2016).

Grain legumes mainly used for feed, such as pea and faba bean, have also experienced an increase of importations, likely due to the high demand of plant proteins, partly influenced by the crisis of the bovine spongiform encephalopathy (BSE) in Europe. Between the 1960s and the 1990s, the pea cultivated area was lower than 50.000 ha. Since then, pea cultivation has significantly increased, reaching 160.500 ha in 2016 (Figure 3). Consequently, pea production has significantly increased, exceeding 290.000 t in 2016 (Figure 4). Castilla-La Mancha (56.765 ha.) and Castilla-Leon (39.647 ha.) are the major regions for pea cultivation in Spain.

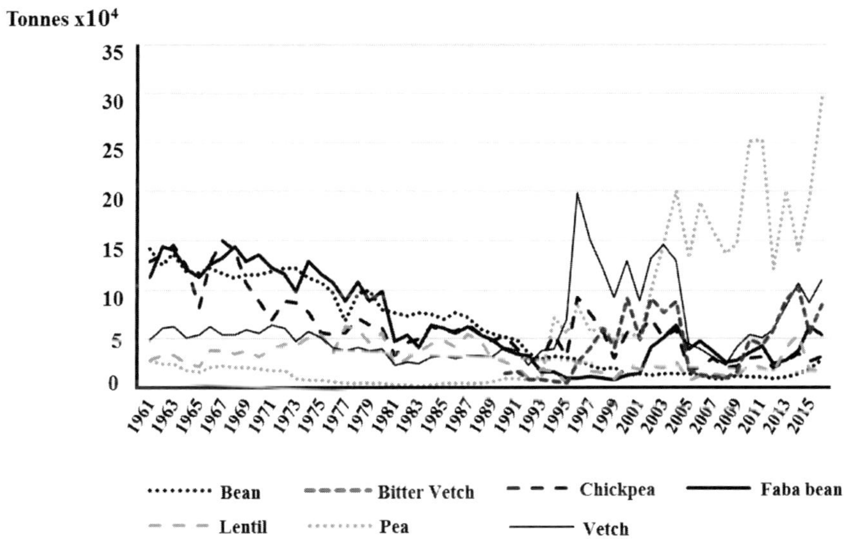

Figure 4. Evolution of production (t) of main grain legumes cultivated in Spain (FAOSTAT 2015; MAPAMA 2016).

Pea is probably the best option to reduce soya imports since it is the only grain legume that has significantly increased the production over the last years in Spain, indicating its high potential and adaptability as well as acceptation. The downside of pea cultivation in Spain is that yields (1.000 kg/ha.) are significantly lower than the average value in Europe (1.800 kg/ha) and in the world (1.700 kg/ha).

Faba bean cultivated area has suffered an extraordinary reduction since the 1960s from 156.200 ha to only 8.900 ha in 1998 (Figure 3). In 2016, the cultivated area reached 45.700 ha, being Andalusia, in the south of Spain, the main faba bean growing region (27.165 ha) (Figure 4). Lupin

(*Lupinus* spp.), despite being an excellent source of protein (28-47%) (Sparvoli et al., 2015), is currently a minor crop in Spain; the cultivated area has decreased from 13.825 to 3.200 ha in the period between 1961 and 2016.

Other grain legumes grown in Spain and mainly used for human consumption are beans, chickpea and lentil. The cultivated area intended for these crops have been reduced over the last 50 years, making Spain a net importer of these grain legumes since domestic production is not sufficient to fulfil internal demands. In 1961, the dry bean cultivation area was higher than 225.000 ha, but today is just 9.300 ha (Figure 3), being mostly cultivated in Castilla-Leon (4.986 ha in 2016). The main reasons for such reduction could be the low competitiveness of Spanish varieties in both national and international markets. It is important to point out the increase of yield over the last decades, from around 600 kg/ha in the 1960s up to 2.000 kg/ha in 2016, being the current annual production 29.300 tonnes (Figure 4). Chickpea has been a traditional crop in Spain for centuries, being the major producer and consumer in Europe. The cultivated area has undergone tremendous fluctuations since 1961, decreasing over the last years (Figure 3). A slight increase in chickpea cultivation was observed during the period between 1995 and 1999; that is explained, at least partially, by the influence of the Common Agricultural Policy (CAP) and the high market prices for first-quality seeds. Chickpea is mainly grown in semiarid regions of the country; these are Andalusia with 16.744 ha and Castilla-Leon with 8.025 ha of the total area of 33.600 ha in 2016. Total production (32.200 tonnes in 2016) does not meet the current national demand, being necessary to import approximately double the domestic production. Around 51.212 tonnes were imported in 2015, mostly from México. There is a clear opportunity for an increase of the chickpea growing area in the near future since this crop is well adapted to our climatic conditions and has a high demand by consumers. Like beans and chickpeas, harvested area and production of lentils has experienced a decline over the last 50 years but Spain still is the leading European producer. The main producer is the region of Castilla-La Mancha with 17.153 ha in 2016. Unfortunately, a cultivated area of 26.600 ha and a total production of 16.900 tonnes is not enough to supply the national demand and more than 60 thousand tonnes were imported in 2013 (FAOSTAT, 2015). In Spain, lentil yield is not very high, with values around 1.000 kg/ha in 2014, far from 1.600 or 1.400 kg/ha obtained in Canada and United States, respectively (FAOSTAT 2015).

Nowadays, vetches and bitter vetch together with peas are the main legumes cultivated in Spain, probably due to policy guidelines. Vetches cultivation area has undergone big fluctuations in terms of harvested area over the last 50 years, reaching a maximum of 300.000 ha in the 90s'. In 2016, 89.900 ha were cultivated with a production of 110.000 tonnes; Castilla-Leon is the main producer region with 67.134 ha. Bitter vetch is mainly grown in Castilla-La Mancha with 52.203 ha out of a total of 71.700 ha, with an average production of 84.800 tonnes.

Alfalfa (*Medicago sativa* L.) is not a grain but a forage legume, mainly used for producing high-value livestock feed. Spain is the second largest alfalfa producer in Europe with more than 260.000 ha of cultivated area. A 58% of this area is located in NE Spain, in the Ebro valley area (Aragon and Catalonia, with 107.000 ha and 42.500 ha, respectively). The cultivated area has been quite stable over time; in 2016, 266.461 ha were cultivated producing 11.105 thousands of tonnes. In the last fifteen years, the sector has been immersed in a process of transformation, which has led to the increase in exports based on dehydrated alfalfa. United Arab Emirates and China were major destinations of Spanish exportations, reaching 75% of the total production in 2016 campaign (AEFA, http://aefa-d.com/). Spain, together with France and northern Italy, are the European countries mostly focused on dehydrated alfalfa. This crop offers new opportunities and could be an alternative for future farming systems.

Overall, the large yield fluctuations in grain legumes and the resulting high variability of returns has probably discourage their cultivation. New policies should be available to change the current situation helping to promote domestic cultivation.

3. The Spanish Network of *Ex Situ* Plant Genetic Resources Collections

According to the Convention on Biological Diversity (CDB), genetic resources are living material that includes genes with current and potential value for humans. Concerning to plants, these resources comprise all our agricultural crops and even some of their wild relatives, weed and breeding material, which could have valuable traits for present or future needs. Since the beginning of the 20th century, the international scientific community has been aware of the need to preserve and characterize these materials, key elements for food security because plant genetic resources are critical to the success of any breeding program (Esquinas-Alcázar, 2005). In the European context, Spain is one of the countries with the highest wild and crop biodiversity. Opportunely, this country promotes and supports systematic activities on plant genetic resources conservation. Significant efforts have been made in the last decades to collect landraces and wild plant populations to ensure their long-term conservation in *ex situ* collections, thus preventing their irreversible lost. Spanish *ex situ* collections of Plant Genetic Resources for Food and Agriculture (PGRFA) are organized in a National Network supported by the National Programme on Conservation and Utilization of Plant Genetic Resources for Food and Agriculture (PCURF). Currently, more than 30 public and associated institutions maintain nearly 90.000 accessions, including both seed and field collections (Figure 5).

Table 1 shows the Spanish institutions involved in the conservation of grain legumes collections. The Spanish National Centre for Plant Genetic Resources belonging to the National Institute for Agricultural and Food Research and Technology (INIA-CRF) plays an important role preventing the loss of genetic diversity contained in these resources. Many of these materials, not cultivated now, would have disappeared if institutions, such as the CRF would not have taken care of their conservation. With these objectives, CRF focuses its activities towards the mid- and long-term seed conservation technologies, field conservation techniques, promotion and control of utilization of active collections, education and awareness on PGRFA conservation, and research in the fields of seed physiology, pathology and technology. In addition to its duty in maintaining active collections of cereals and grain legumes and conserving safety duplicates of all seed collections of the network, CRF is responsible for the documentation of the whole Network (De la Rosa et al, 2014). Following this mandate, CRF developed in 2000 the National Inventory (NI) of PGRFA, which is regularly updated and available on-line[1], in both Spanish and English.

Figure 5. Map institutions participating in the National Network of *ex situ* PGRFA collections.

[1]http://wwwx.inia.es/inventarionacional/.

Table 1. Spanish Institutions involved in grain legumes conservation within the National Program of Plant Genetic Resources for Food and Agriculture (PGRFA)

Institution	Province	WIEWS[1] code
Germplasm Bank César Gómez-Campo (Polytechnic University of Madrid)	Madrid	ESP003
National Center of Phytogenetic Resources (INIA)[2]	Madrid	ESP004
The Biological Mission of Galicia (CSIC)[3]	Pontevedra	ESP009
Center of Agricultural Research Finca La Orden-Valdesequera	Badajoz	ESP010
Neiker-Tecnalia. Basque Institute for Agricultural Research and Development	Álava	ESP016
Germplasm Bank of the Polytechnic University of Valencia	Valencia	ESP026
Center of Research and Food Technology of Aragón	Zaragoza	ESP027
Regional Service of Research and Food Development	Asturias	ESP032
Center Alameda del Obispo (IFAPA)[4]	Córdoba	ESP046
School of Engineering of Biosystems and food technology of Barcelona (Polytechnic University of Catalonia)	Barcelona	ESP061
Zamadueñas Research Center (ITACYL)[5]	Valladolid	ESP109
La Orotava Acclimatisation Gardens (ICIA)[6]	Tenerife	ESP117
Agroforestry Research Center of Albaladejito (Castilla-La Mancha)	Cuenca	ESP124
Centre for the Conservation of Agricultural Biodiversity in Tenerife	Tenerife	ESP172
Bank of local varieties, Madrid (IMIDRA)[7]	Madrid	ESP198
Institute of Agricultural and Fisheries Research and Training (Illes Balears)	Palma	ESP200

[1]VIEWS: World Information and Early Warning System on Plant Genetic Resources for Food and Agriculture (PGRFA) has been established by FAO as a world-wide dynamic mechanism to foster information exchange among Member Countries and as an instrument for the periodic assessment of the State of the World's PGRFA.

[2]INIA: National Institute for Agronomic Research.

[3]CSIC: The Spanish National Research Council.

[4]IFAPA: Andalusian Institute of Agricultural and Fisheries Research and Training.

[5]ITACYL: Technological Institute of Castilla-Leon.

[6]ICIA: Canarian Institute of Agricultural Research.

[7]IMIDRA: The Madrid Institute for Research and Rural Development in Food and Agriculture.

4. THE SPANISH COLLECTIONS OF GRAIN LEGUMES

Many grain legume crops (common bean, pea, chickpea, lentil, faba bean, lupin and vetch, among others), are well adapted to European and Spanish agro-ecological conditions but their cultivation have decreased since the middle of the last century. Substantial changes in agricultural practices and consumption habits mainly happened in the first half of the 20th century whereas massive imports of soya at low prices for feed occurred in the second half; as a result, an important genetic erosion of this heritage has taken place. Spanish grain legumes have been deeply affected; some species have disappeared from the agricultural landscape and others have lost a great part of their intraspecific variability, due to the substitution of landraces by commercial varieties. In the National Genebank Network of PGRFA, a great diversity of grain legumes is conserved, including a total of 14.441 accessions identified with passport data. In fact, this group is the second largest, after the group of vegetable crops, in the NI of PGRFA and it reflects the importance of grain legumes in the Spanish traditional agriculture. The distribution by genus is shown in Figure 6. Regarding the type of material, most of these accessions correspond to Spanish local varieties or landraces donated by farmers, crop wild relatives (CWR) as well as outdated commercial varieties.

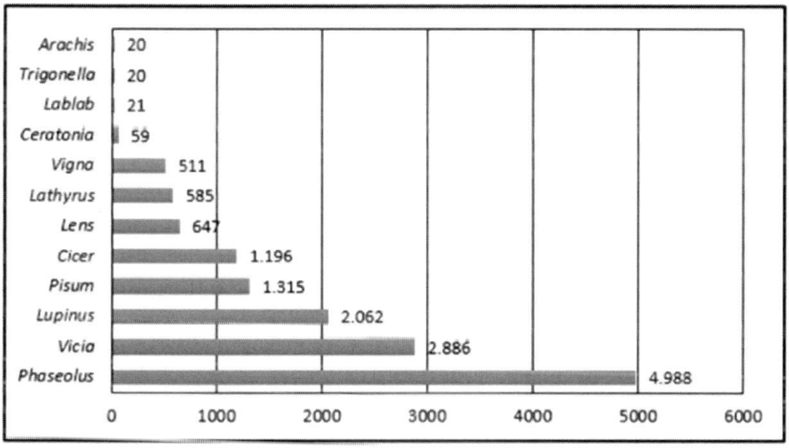

Figure 6. Distribution by genus of grain legumes collection in the NI.

Systematic collecting activities began in Western Spain in the late 1970s supported by FAO and the World Bank. Since then, a very large number of grain legumes landraces have been collected. Thus, during the period 1978-2014, collecting expeditions have been carried out in 42 Spanish provinces. The result has been the collection of 4.687 accessions, whose distribution by genus is shown in Figure 7. The marginal agricultural areas are the places where is easier to find the most valuable plant genetic resources, the landraces. Traditional farmers maintain varieties inherited from their ancestors because they are aware of the importance of safeguarding this heritage and because their sensorial characteristics are preferred.

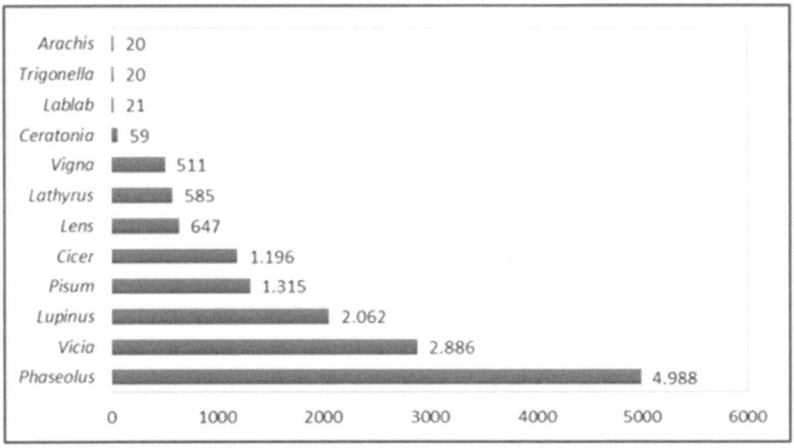

Figure 7. Distribution by genus of grain legumes accessions collected by the CRF during the period 1978-2014.

The situation of the Spanish grain legumes collection is detailed below:

- **Common bean.** CRF conserves the most important collection of this specie, with 3.590 accessions. Other large bean collections are those of MBG-CSIC (Misión Biológica de Galicia), SERIDA (Regional Service of Research and Food Development), ESAB-UPC (Barcelona School of Agricultural Engineering-Polytechnic

University of Catalonia), Neiker and León University. All of them include working materials from international centres or comprehensively collected in their influence zone.

- **Pea.** In addition to the 685 pea accessions of the CRF, ITACYL (Technological Institute of Castilla-Leon) holds a great collection from international genebanks that have been used in breeding programmes by this institution.
- **Chickpea.** CRF preserves the most important collection (843 accessions) but should be also mentioned IFAPA (Andalusian Institute of Agricultural and Fisheries Research and Training) where a secondary active collection is maintained. ITACYL and CCBAT (Center for Conservation of Agricultural Biodiversity of Tenerife) maintain smaller but important collections from very different eco-geographical regions.
- **Lentil.** Long-term lentil collection (594 accessions) is preserved at the CRF, but the mid-term collection is managed by the CIAF (Agroforestry Research Center of Albaladejito). Most of the accessions belong to *L. culinaris*, but there are also 48 *L. nigricans,* 14 *L. lamottei* and 9 *L. ervoides* accessions. At present, CRF is collecting wild lentil populations in order to increase the number of accessions of these species, scarcely represented in genebanks.
- **Faba bean and vetch.** *Vicia* genus is the second grain legume collection in terms of the number of accessions. A total of 2.999 accessions are conserved in the CRF, 1.375 belonging to faba bean. This collection was established by IFAPA with genetic resources from Spain and from international genebanks, mainly ICARDA (International Center for Agricultural Research in the Dry Areas), to meet its needs in breeding programmes. *V. sativa* group (*sativa, amphicarpa, macrocarpa* and *nigra*) includes 995 accessions, most of them of Spanish origin. Part of vetch collection (523 accessions) has been the subject of morphological, biochemical and molecular (ISSR) characterization; after data analysis, a preliminary core collection has been established. There

are other *Vicia* species with a reduced number of accessions corresponding to *V. ervilia* (L.) Willd (271), *V. articulata* Hormen (112), *V. villosa* L. (115) and *V. narbonensis* L. (26). These species were an important element in former agricultural practices and could play again an important role in crop rotation and in sustainable agriculture.

- **Lupin.** Spain is a lupin diversification centre; therefore, collection and conservation of species of this genus in Spain were a mandate from international institutions in the seventies of last century. The result is a collection of 2.158 accessions, mainly from Spain and Portugal, comprising 12 species. Most of the accessions are from cultivated species: *L. albus* (754), *L. angustifolius* (703), *L. luteus* (371), *L. gredensis* (157) and *L. hispanicus* (70). The mid-term collection is preserved by CICYTEX (Center for Scientific and Technological Research of Extremadura) while a long-term collection is conserved by CRF.
- ***Lathyrus.*** This collection includes 477 accessions, mainly integrated by grass pea (*Lathyrus sativus* L., 265) and red pea (*Lathyrus cicera* L., 231); *L. amphicarpos, L. aphaca, L. clymenum, L. tingitanus, L. ochrus* and *L. hirsutus* are also well represented in the collection.
- **Cowpea (*Vigna unguiculata* L. Walp).** CRF conserves 493 accessions, mostly provided by other genebanks. Only 126 of the total come from Spain.

Passport and characterization data of all the Spanish grain legumes collections are available in the NI (http://wwwx.inia.es/inventarion acional/) and CRF (http://wwwx.inia.es/coleccionescrf/) websites.

Grain legume seeds are predominantly orthodox, and retain a high viability level in low moisture and temperature conditions; consequently, their conservation status at CRF facilities is very good for most of the species. Nevertheless, germination level is usually tested to detect viability falls below a threshold. When this happens, the accession is regenerated, following the most appropriate procedure (FAO, 2013). In cross-pollinated species, this task may be complicated and suitable control pollination measures must be taken.

One of the reasons that explain the low utilization rates of the germplasm collections is the scarcity of characterization and evaluation data. In response to this pitfall, the establishment of core collections enables to focus characterization and evaluation efforts on a smaller number of accessions and increase their use (Odong et al., 2013). Different type of descriptors may be used to set a core collection: morphological, molecular, eco-geographical, etc. At CRF, two grain legumes core collections have been established using mainly morphological and agronomic variables: a bean core collection, which includes 202 accessions, and a vetch core collection with 54 accessions. Eco-geographical information has been proved to be useful in the selection of accessions of the bean core collection, so this approach could be used in other species (Parra-Quijano et al., 2011). It is advisable to establish and expand the germplasm collections, including CWR. Climatic change along with habitat loss and fragmentation by human activity, it is causing an important loss of genetic diversity in this group of species, so their collection and conservation is urgent. However, actions should be taken to allow grain legumes landraces be present in agrarian systems, now and in the future.

5. Impact of the Common Agricultural Policy (CAP) in the Spanish Legume Crop Production

The EU Common Agricultural Policy (CAP) was developed in the early sixties to protect agriculture throughout the EU by influencing prices, output and farmers′ incomes. Being one of the first common policies for several European countries, it aimed at fostering food security and stabilizing agricultural markets by introducing product and price support. This was achieved largely by decoupling farm subsidies from production activities and output. As the environmental costs of production became more apparent, various reforms of the CAP emphasized environmental and rural development measures and greater exposure to market conditions to halt over-production.

Protein crops have been subject over a long period of time to special support measures to overcome competitive disadvantages compared with cereals. Among the CAP measures for supporting legumes, it deserves special mention the Council Regulation (EEC) No 1119/78, laying down special measures for "protein crops", such as pea, field bean and lupine, grown for livestock in 1978 and for the same crops grown for food uses in 1982. The objective was to compensate farmers for their income losses caused by our dependency on soya bean import for the animal feed industry. Spain was able to benefit from these payments after joining the European Common Market in 1986. However, the new context did not produce an appreciable rise in the pea cultivation area until the mid-1990s and the effect in the case of faba bean was rather limited.

The CAP has undergone significant modifications since early 1990s, with the aim of reducing market distortions or making them more acceptable at the international level. To alleviate competition between legume species, a support to traditional southern European grain legumes such as chickpea, lentil (both for food) and vetches (for feed) was introduced in 1989 in the approach of uniform area payments that were continued until 2005/06. The payment amount was doubled in 1995, having and immediate response with more than 700,000 ha of grain legumes cultivation, a record for Spain. Nevertheless, the growing demands of seeds could not be supplied with domestic production only, leading to a massive importation of varieties that, in most cases, were not well adapted to our particular conditions. The low productions thus achieved discouraged farmers and declined sharply the acreage. This trend was accentuated from 2006, when the single payment scheme began to be applied in Spain and was only partially reversed in 2009. Both bitter vetches and chickpeas followed a similar trend to that of the vetches, while the lentils experienced a decline in their cultivated areas.

The profound amendment of 2003 was marked by decoupling some direct payments from production to give clearer market signals to farmers.

The rationale behind the introduction of decoupled payments was that, if direct payments to farmers were not related to production, they would not be distorting. However, the payments were linked to the fulfillment of certain requirements as the obligation of farmers to conserve the environment and keep their land in "good agricultural conditions." The inexperience and uncertainty in this type of policies drove to a brutal fall in the legumes cultivated area (to 200.000 ha. in 2008), being legumes the most damaged group of crops in this reform (Torres et al., 2015).

The next modification in 2008 was the Health Check which introduced short-term adjustments in European regulations. It was focused towards the decoupling of most direct aids to the production with some exceptions, as occurred with protein crops, which were allowed to have a transitional period until 2012. The decline of legume cultivation was so extreme in 2008 that two programmes were launched in Spain to reverse such situation: The National Program for the Promotion of Crop Rotation in Drylands and the National Program for the Quality of Pulses. They met partially their goals, rising from 311.000 cultivated ha in 2009 to almost 450.000 ha in 2010 and more than half a million in 2011. These figures, although higher than those of 2008, remain very far from the ideal surface that should cover this group of crops to meet the environmental needs and supply of raw materials for the internal market.

The most recent CAP reform proposal was decided in 2013, published in Spain in 2014 (BOE nº. 307, December 20, 2014) and implemented in 2015. An important objective was the maintenance of farming as a viable economic sector, which also helps to take care of the environment as well as maintaining rural communities and landscapes. It was proposed a system of two pillars: Pillar 1 support included both direct payments to farmers and market management measures, but also comprised a "greening" component (the main source of income for most farmers) together with a number of coupled subsidies to legumes in order to promote protein crops for animal feeding; Pillar 2 support focused on improving the structural and environmental performance of agriculture.

The "greening" component means that 30% of direct payments would be paid to farmers only if they comply with three environmentally-friendly farming practices:

- Crop diversification: in farm holdings with more than 30 ha of arable land, at least three crops shall be cultivated, covering each of them an area between 5% and 70% of the arable land.
- Maintaining permanent grassland.
- Establishment of ecological focus area applying to 7% of the eligible area, excluding permanent grassland.

The current CAP reform proposal includes two specific elements relevant to protein crops:

- Inclusion of nitrogen-fixing crops (i.e., legumes) in ecological focus areas;
- A call from the European Parliament for the Commission to prepare a strategic plan for the production of oil-protein crops and legumes to be grown under the common agricultural policy.

The Spanish authorities have decided to apply the Basic Payment (which replaces the Single Payment) on a regional level with a differentiation that takes into account institutional, agronomic and economic criteria (AEL, 2017). In addition, Spain's authorities have decided to earmark 12.1% from its Direct Payment for voluntary coupled support (in which protein crops are included). Thus, since 2015, the PAC offers two types of aids associated to legumes, very similar to those described so far. The first is associated to "protein crops," created with the aim of promoting protein crops for animal feeding. According to the Spanish Agricultural Ministry, the amount can be around 60 €/ ha for legumes, in both irrigation and rainfed conditions producing over 2.000 kg/ha. The crops species considered are pea, faba bean, sweet lupine, vetch, bitter vetch, bard vetch, chickling vetch, grass pea, fenugreek, cow vetch, narvon vetch and alfalfa (rainfed only), sainfoin and sulla, together with some oilseeds. The basic requirements to accomplish are:

1. The use of varieties registered in the Common Catalog of Varieties of the EU.
2. Proof of sale or supply to third parties or self-consumption on the farm.
3. To carry out the agricultural work that ensures the normal development of the crop.
4. To maintain it in the field until the milky maturity stage of the grain, in the case of protein crops and legumes for grain, and until the beginning of flowering, in the case of protein crops and annual forage legumes.

The EU also provides a number of measures to help producers build on the high quality reputation of European products to sustain competitiveness and profitability. Thus, the second aid is associated to "Quality Pulses" and can be applied to farmers growing lentils, chickpeas or beans varieties classified as a Protected Designation of Origin (PDO), a Protected Geographical Indication (PGI) or an organic farming product. This aid has a budget of 1 million euros, and the Spanish Ministry estimates an aid of 100 €/ha for a maximum of 10,000 ha in the whole national territory. The total EU budget fixed for the period from 2014-2020 is of EUR 408.31 billion with EUR 308.73 billion intended for direct payments and market measures (First Pillar) and EUR 99.58 billion for Rural Development (Second Pillar). In the period to 2020, the new CAP will invest almost EUR 45 billion (near 11% of the total) in Spain's farming sector and rural areas.

With nearly 506,000 km^2 (50% agricultural) and 46 million people (7.3% of the population living in rural areas), Spain has a dynamic agri-food industry and a highly heterogeneous agriculture, being among the main economic sectors in terms of turnover and exports. The new reform of the CAP has raised the interest in Spain for arable crops (producers, cooperatives and seed and agricultural supplies companies) and has boosted the interest for growing legume crops. Despite this reactivation, the growth experienced in Europe has not been so noticeable in Spain. To farmers, who make long-term decisions, legumes have already found their place in rotations, mostly because of the increased productivity in the next crop than for their profitability during the growing season. The crop diversification requirement to receive the Green Payment and the second aid associated to "Quality Pulses" classified as PDO, PGI or organic farming products has placed the legumes in the spotlight to be considered an interesting crop alternative. Nevertheless, it should be remarked that the "Quality Pulses" aid discards most of the lentils, beans and chickpeas produced in Spain as only about 5,000 ha grouped under the denominations "Faba Asturiana," "Mongeta del Ganxet Vallés-Maresme," " Fesols de Santa Pau," "Judía del Barco de Ávila," "Alubia de la Bañeza Pinta y Blanca," "Faba de Lourenzá," "Lenteja de la Armuña," "Lenteja Pardina de Tierra de Campos," "Garbanzo de Fuentesaúco" and "Garbanzo de Escacena" met these requirements (Figure 8). Thus, for farmers involved under these denominations and for those who practice organic farming, the "Quality Pulses" aid will increase their profitability and, therefore, the declared cultivated areas. In contrast, for the rest of producers of quality legumes for human consumption (lentils, chickpeas and beans) which do not meet the above requirements (more than 80% of the total), the lack of a coupled aid is a frustration that can lead even to a decrease of the surfaces devoted to these crop species. Future revisions of the CAP should consider a similar support to protein crops for animal feeding, since Spain is deficient in both quality and protein crop production.

Granting subsidies to support these crops are necessary, but not enough to ensure the long-term health of the legume sector. The increase in cultivated area should be supported by the use of certified seeds, resistance to pests and diseases, adapted to the different agro-climatic conditions and providing good yields. On the other hand, it is uncertain to subordinate the profitability of legume crops to European subsidies, considering the political insecurity that will exist in the European Union itself in the next years. Besides, it would also be important to carry out permanent campaigns to promote legume consumption among the population and disseminate the results of the growing numbers of studies showing their beneficial effects on human health. This initiative, however, would be useless unless the campaigns were more focused in the consumption of high quality domestic legumes, supporting national PDO and PGI. The fact that two-thirds of the legumes consumed in Spain are imported indicates that there is a wide margin to increase their cultivated area and guarantee food security.

Figure 8. Quality Pulses in Spain.

The existing dependence on soybean imports for animal feeding in Spain places the livestock sector in a clear situation of vulnerability. Although several soybean replacement studies have been developed showing very positive results, large-scale alternatives including native legumes in the feed formulation should not be improvised. This conversion would require promoting cultivation of native legume species which, in many cases, present a higher rusticity and lower demands and meet the nutritional requirements of the feed producing sector. Overall, the proposed measures should place legumes cultivation and consumption in the position they deserve in the future farming and food systems.

6. Dietary Habits and Legume Consumption in Spain

In Spain, dietary habits have greatly changed over the last years as standard of living has improved. These changes have not been necessarily positive since people have abandoned the Mediterranean lifestyle and diet and are moving to a more Occidentalized way of living, meaning among other things consuming less legumes, fruits and vegetables and more animal-protein (Valera, 2014).

Spain uses a food pyramid with dietary recommendations in line with the traditional Mediterranean diet. In this sense, pulses combined with cereals, fruits and vegetables play a major role in a balance diet. Although Spain is one of the European countries where more pulses are consumed (FAOSTAT, 2015), there has been a great decline in consumption, higher than 60%, over the last decades (Figure 9). The volume of pulses consumed in Spain during 2014 was 140.888 tonnes, with an annual per capita consumption of 3.1 kg (MAGRAMA, 2015). Nowadays, chickpeas (1.25kg/capita), beans (0.93 kg/capita) and lentils (0.93/capita) are the legumes more consumed in Spanish households. Although the tradition of pulse consumption remains strong, it has experienced a decline in all income groups (rural areas and urban areas), being this drop more significant in higher socio-economic groups (urban areas).

There is an increasing interest in promoting legume consumption particularly since several studies have shown that legumes have potential health benefits in humans such as reduced risk to suffer cardiovascular disease (CVD), obesity and weight management, inflammatory disorders and cancer (Delgado-Andrade et al., 2016; Clemente and Olias, 2017). For these reasons, increasing legume intake is gaining interest in the Spanish population, being highly recommended by different health and nutrition organizations. The composition of grain legumes makes them exceptional ingredients to improve the nutritional quality of food products, conferring low glycemic index, a good antioxidant profile as well as a great capacity to hold water and to absorb fat. Scientists and the food industry are currently working together to design new food products based in grain legumes aiming to restore the consumption of this important source of proteins, minerals, vitamins and fiber together with other health promoting components in the diet. Other reason for taking into account the use of grain legumes in the food industry is the fact that these are gluten-free, making them an excellent option for people with special needs such as celiacs whose only treatment is to avoid gluten. In this sense, flour from different legume sources is currently being used to increase the nutritional quality of gluten-free products otherwise very poor in nutrients (Foschia et al., 2017). These legume-flours are also being used either on its own or combined with cereal flour in the design of new healthy snacks, fortifying the final products in its amino acid profile and mineral content (Rababah et al., 2012). All these innovations urges the consumption of healthy and nutritious diets and, indirectly, to increase the intake of legumes, due to the advantages that offer new foods to the traditional (Vaz Patto et al., 2015).

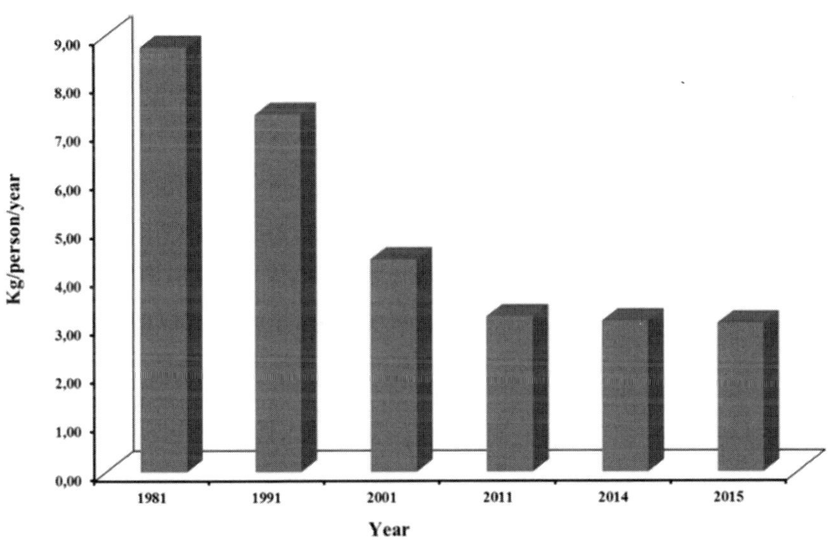

Figure 9. Evolution of grain legume consumption in Spain over the last decades (kg/person/year). Sources: Valera et al. 2014, MAPAMA 2016.

In order to promote healthy diets, different recommendations have been published by national and international administrations and agencies. They are usually expressed in terms of portions/week, being a portion 60-80 grams of dry uncooked legumes or 150-200 grams of cooked legumes. The Spanish Nutrition Foundation (www.fen.org.es) encourages a weekly intake of 3-4 portions of legumes (SENC y SEMFYC, 2007), being the current intake 1.4 portions per week in the adult population. In Spain, legumes consumption declined a 3.3% in 2015 with respect to 2014. Latest reported data reflects a steady decline compared to the same period of the previous year (http://www.agro-alimentarias.coop/ficheros/doc/03181.pdf). Taking into account this trend, an urgent plan of action should be undertaken in order to promote regular legume consumption.

References

AEL (Spanish Association of Legumes). Position of the AEL on the CAP reform. Meeting with the Agriculture Ministry. Madrid, March 27-28th, 2017.

Clemente, A. and Olias, R. (2017). Beneficial effects of legumes in gut health. *Current Opinion in Food Science*, 14: 32-36.

De la Rosa, L., Aguiriano, E. and Fajardo, J. (2014). The documentation of *ex situ* collections of Plant Genetic Resources in Spain. Available at: http://www.ecpgr.cgiar.org/fileadmin/templates/ecpgr.org/upload/NW_and_WG_UPLOADS/Doc_Info_2014/Spanish_PGR_documentation.pdf.

Delgado-Andrade, C., Olías, R., Jiménez-López, J. and Clemente, A. (2016). Aspectos de las legumbres nutricionales y beneficiosos para la salud humana. *Arbor*, 192 (779): a313.

Esquinas-Alcázar, J. (2005). Protecting crop genetic diversity for food security: political, ethical and technical challenges. *Nature Reviews Genetics,* 6(12): 946-953.

FAO. (2013). Genebank Standards for Plant Genetic Resources for Food and Agriculture. Rome.

FAOSTAT. (2015). Available in: http://faostat3.fao.org/ download/ Q/QC/S.

Foschia, M., Horstmann, S.W., Arendt, E.K. and Zannini, E. (2017). Legumes as functional ingredients in gluten-free bakery and pasta products. *Annual Review Food Science Technology*, 8:75-96.

MAPAMA (2016). Available in: http://www.mapama.gob.es/es/ agricultura/temas/producciones-agricolas/cultivos-herbaceos/leguminosas-y-oleaginosas/ and http://www.mapama.gob.es/imagenes/es/ informeconsumoalimentacion2015.

Odong, T.L., Jansen, J., Van Eeuwijk, F.A. and Van Hintum, T.J.L. (2013). Quality of core collections for effective utilisation of genetic resources review, discussion and interpretation. *Theoretical and Applied Genetics*, 126(2): 289-305.

Parra-Quijano, M., Iriondo, J.M., Torres, E. and De la Rosa, L. (2011). Evaluation and validation of ecogeographical core collections using phenotypic data. *Crop Science*, 51 (2) : 694-703.

Rababah, T.M., Brewer, S., Yang, W., Al-Mahasneh, M., Al-U'Datt, M., Rababa, S. and Ereifej, K. (2012). Physicochemical properties of fortified corn chips with broad bean flour, chickpea flour or isolated soy protein: fortification of corn chips. *Journal of Food Quality,* 35 (3): 200-206.

Sparvoli, F., Bollini, R., and Cominelli, E. (2015). Nutritional value. *In*: Antonio M. de Ron (ed.) *Grain Legumes*. Springer, pp 291-325.

Torres, A.M., Rubio, J., Millán, T., Eraso, E. and Solís, I. (2015). El papel de las leguminosas en la PAC: realidad y perspectivas. Revista Ambienta nº 112, September (www.revistaambienta.es).

Varela, G. (2014). La dieta española, fortaleza y debilidades. *Nutrición Clínica en Medicina,* 8: 109-120.

Vaz Patto, M.C., Amarowicz, R., Aryee, A.N.A., Boye, J.I., Chung, H.J., Martín-Cabrejas, M.A. and Domoney, C. (2015). Achievements and Challenges in Improving the Nutritional Quality of Food Legumes. *Critical Reviews in Plant Sciences,* 34: 105-143.

In: Legumes for Global Food Security ISBN: 978-1-53612-265-7
Editors: Alfonso Clemente et al.

Chapter 2

LEGUMES AND CLIMATE CHANGE

***Gemma Molero*[1] *and Iker Aranjuelo*[2,*]**
[1]CIMMYT (International Maize and Wheat Improvement Center), Mexico D.F., Mexico
[2]Instituto de Agrobiotecnología (IdAB), CSIC-Universidad Pública de Navarra-Gobierno de Navarra, Mutilva Baja, Navarra

ABSTRACT

Legumes are relevant at agronomical and economical level as their high protein content (between 26 - 30%) is of primary importance for human food and animal feed. In addition, their ability to establish symbiotic relationships with soil bacteria allows them to assimilate atmospheric N_2 reducing the use of nitrogen fertilizers. Symbiotic nitrogen fixation is highly sensitive to water availability and ambient temperature determining legume development. The current book chapter provides a compilation of results from existing literature on the effects of several abiotic stress (drought, high temperature and CO_2 concentration [CO_2]) conditions on plant growth, nodule performance and N_2 fixation. Available information confirms that increasing [CO_2], water scarcity and high temperature strongly affects N_2 fixation and plant growth. The

* Correspondence Author E-mail: iker.aranjuelo@csic.es.

effects of water and temperature stress on plant and nodule includes not only reduction on biomass production, photosynthetic rates but a decreases in leaf and nodule functioning. In the other hand, long-term high [CO_2] studies remark that while plant growth is stimulated as a first instance, photosynthesis down-regulation (associated with a C/N imbalance) may limit leaf N demand and hence, nodule activity.

1. Relevance of Legumes

In the last decades, the amount of synthetic nitrogen (N) applied to crops has risen dramatically (Mulvaney et al., 2009), resulting in significant increases in yield but with considerable impacts on the environment worldwide. Traditional breeding strategies aiming to improve nitrogen use efficiency (NUE) in crop plants have experienced a plateau, where increases in N applied do not result in yield improvements. This indicates that new solutions are needed to increase yields while maintaining, or preferably decreasing, applied N to obtain the estimated attainable yields of these plants under specific nutrient regimes.

Legumes represent the second worldwide major crop (preceded by grasses) and cover the 12-15% of world cultivated surface (Alexandratos and Bruinsma, 2012). Those plants represent an important source of protein and calories for human and animals (Rogers et al., 2009). In addition, and compared with other cultivars, N_2 fixing plants represent a particular plant group thanks to their capacity of fertilizing soils through the fixation of atmospheric N_2. The fact that these plants have the capacity symbiotic N_2-fixing bacteria, gives them access to a large reservoir of natural N_2 in the atmosphere. This symbiotic relationship explains the fact that that those plants act as natural fertilizers releasing the N to the soil. This release is due to exudation and decomposition of roots and nodules.

The importance of legumes also lies in their use in human food (grain legumes) and feeding of livestock (grain legumes and forage). Legumes contain, approximately, double the amount of proteins found in whole grain of cereals such as the wheat. For this reason, legumes might represent a relevant source of protein. Therefore legumes are globally important in second world crops (preceded by grasses) and represent between 12 and 15% of the Earth's cultivated land (Graham and Vance, 2003).

2. Changes in Current and Near Future Environmental Conditions

The Intergovernmental Panel on Climate Change (IPCC), created in 1988, by scientists from all over the world, was developed with the goal of providing objective and truthful information on climate change and other related aspects. The IPCC has drawn up scenarios of possible future emissions elaborated from predictions of the World Bank and the United Nations, taking into account the global demographic and economic growth. In relation to the concentration of atmospheric CO_2, [CO_2], the IPCC indicates that, fruit of the human activity, the [CO_2] in the atmosphere has undergone a continuous increase since the beginning of Industrial Revolution (1750). During the previous thousand years to Industrial Revolution, the concentration of [CO_2] in the atmosphere was stable; around 270 μmol mol^{-1}. Today the average [CO_2] of the atmospheres is around 400 μmol mol^{-1}. According to the IPCC, it is expected that [CO_2] will rise up to 550 μmol mol^{-1} by 2050, reaching 700 μmol mol^{-1} by the end of century.

Models predicted for the following decades, forecast an increase of the atmospheric temperature of between 1-5.8° C. The thermal increase will entail an increase in evapotranspiration rates, In an ecosystem where the annual potential evapotranspiration is often twice as high as the annual precipitation. The reports also postulates to generalized decline in annual totals precipitations. The most critical areas will be semi-arid regions, where precipitation can be reduced up to 50% above the current potential. The impact of climate change on water resources not only from inputs comes from the hydrological cycle, but from the available water resources systems and the efficiency with which they are used.

3. Impact of Climate Change in Legume Development

Climate change effect on plant growth is of major concern because, as it has been explained in more detail above, most of the recent global climate models predict an increase in temperature and decrease in water availability by the following decades. Therefore, global agricultural production will be profoundly impacted. The adverse effects of global warming and water shortage on plant development have been highlighted for crop production in Europe (Ciais et al., 2005; Oury et al. 2012). As an example, the precipitation deficit of up to 300 mm $year^{-1}$ (50% below the average) together with the 6 °C increase in average temperature experienced by Europe during July 2003, reduced European plant production to the lowest values recorded in the last 100 years. These rapid changes portend critical uncertainty over both carbon uptake by terrestrial ecosystems and food production. For this purpose, it is essential to identify target legumes cultivars better adapted to predicted future climates.

3.1. Water Stress and Elevated Temperature Effect

Although legumes such as alfalfa have been described to be better adapted to water stress and elevated temperature conditions, predicted rising spring and summer temperatures may lead to stagnating yields. Various and numerous processes of the plant can be altered by water stress and high temperature. Responses may be transient or may involve changes in gene expression. Both stresses can be considered a complex syndrome, composed of a large number of processes, some of which are deleterious and others are adaptive (Chaves et al., 2002).

One of the first effects of water deficit and high temperature on plants is the inhibition of cell elongation and, therefore, of growth (Aranjuelo et al., 2009). Leaf area variation is one of the earliest macroscopic responses in plants subject to water deficit. In case of water stressed plants root growth usually stops later and is less inhibited than aerial part leading to an increase in root / shoot ratio. Plant water status is also negatively affected by drought and high temperature conditions (Lawlor and Tezara 2009; Aranjuelo et al., 2009). The accumulation of osmolytes in the plant cells results in a decrease in osmotic potential (Ψs) allowing the maintenance of water absorption and turgidity pressure. Under conditions of water and high temperature stress this can contribute to the maintenance of various physiological processes (Figure 1) such as stomatal opening, photosynthesis and cell expansion (Aranjuelo et al., 2011). In addition, in many cases these osmolytes (known as 'osmoprotectors') can stabilize cell membranes by protecting macromolecules such as proteins and lipids, which would help to preserve the subcellular structures of the damage caused by cell dehydration in crops such as legumes (Aranjuelo et al., 2011).

The inhibitory effect of high temperature and drought on legume gas exchange parameters is a broadly characterized fact (Chaves et al., 2002). This limitation of photosynthesis may be due to (i) stomatal factors and / or to (ii) non-stomatal or metabolic factors (Figure 1). Such causes of inhibition can affect the assimilation of CO_2 singly or jointly. There are numerous evidences that emphasize the stomatal closure as a target event in conditions of water stress and elevate temperature conditions (Flexas et al., 2002). The stomatal closure is due to a decrease in cellular turgor and/or water potentialor to a low relative humidity in the environment. Stomatal responses are usually more related to the water content of the soil than to the water content of the leaf. The phenomenon is related to the increase of the endogenous levels of abscisic acid (ABA) in leaves that mainly comes from the roots (Flexas and Medrano 2002). The result of the stomatal closure is the reduction of the transpiratory rate and, consequently, of the water consumed by the plant. The decrease in stomatal conductance (and the parallel decrease in intercellular [CO_2] or C_i concentration values) leads to a fall in CO_2 assimilation and various associated effects such as ATP accumulation and reducing power, susceptibility to photoinhibition and/or photooxidation, increased photorespiration, formation of reactive oxygen species (ROS), etc.

Although stomatal closure is considered the main factor limiting photosynthetic activity under moderate water and high temperature stress conditions (Aranjuelo et al., 2007), when the stress is more severe metabolic limitations occur (Medrano et al., 2002). Among the causes associated with the possible existence of metabolic limitations of photosynthesis derived from water stress, we can cite: (i) limitation by activity of Rubisco; (ii) regeneration of Ribulose-1,5-biphosphate (RuBP) in the Calvin Cycle (PCR); (iii) limitation of the availability of ATP and NADPH in the PCR and (iv) decrease in electronic transport and transport of protons at the thylakoidal level (Lawlor and Cornic 2002).

Biological nitrogen fixation (BNF) is especially sensitive to water and high temperature stress with a rapid inhibition of nitrogenase activity (Zahran, 1999) (Figure 1). Several studies have shown that N_2 fixation is more sensitive to water stress than leaf gas exchange (Serraj et al., 1999; Aranjuelo et al., 2008). Water and temperature derived inhibition has been associated with different mechanisms such as (i) decreased permeability of O_2; (ii) reduction of the carbon compounds to the bacteroid; (iii) accumulation of nitrogen compounds (which may inhibit nitrogenase activity by negative feedback) and (iv) oxidative stress. The relative importance of these mechanisms is currently a topic of debate and may be related to the structural and functional variability existing in the symbiosis between the rhizobium and the legume (Schulze, 2004).

3.2. Elevate [CO_2] Effect

Plants are photosynthetic organisms that use atmospheric [CO_2] as the main substrate of photosynthesis and, therefore, as the main source of growth. For C_3 plants, ie for most photosynthesizing organisms, atmospheric [CO_2] concentration is suboptimal. Therefore, it is logical to expect a "fertilizer" effect associated with the increase of atmospheric [CO_2] levels, which will result in greater plant growth (Figure 2). Most experiments analyzing the climate change effect on plant growth have considered the effect of [CO_2] or [CO_2] ×temperature under optimal water conditions. However, as mentioned above, it should be considered that circulation models predict drier conditions to an increase in temperature and water deficit (IPCC, 20015). It has been observed that the effect of increasing [CO_2] on the growth of alfalfa plants depends on the temperature to which it is applied and water availability (De Luis, 1999; Aranjuelo et al., 2007; Erice et al., 2011). High [CO_2] applied at 25° C (optimum growth temperature for alfalfa) stimulated plant growth by 61%, while when applied at 28° C (supraoptimal temperature) the increase was considerably lower (15%) (De Luis, 1999). Rawson et al. (1995), in experiments carried out in thermal gradient tunnels, also observed that the effect of [CO_2] depended on the temperature of growth. When the experiment was performed in winter, with an average temperature of 12.5° C, [CO_2] stimulated growth by 7%. An increase of 2° C did not change the effect of [CO_2]. However, in summer and with an average temperature of 20° C, the increase was 34%.

While the higher [CO_2] has been described to increase growth of legumes (Rogers et al., 2009), initial photosynthetic stimulation results in a decrease in the carboxylation efficiency associated with loss of photosynthetic capacity (Long et al., 2004). This phenomenon is better known as "photosynthetic acclimation." Different hypotheses have been proposed to explain this phenomenon. A first hypothesis would associate this phenomenon with the increase in the source/sink relationship of plants. This imbalance would affect the decrease in the content and/or activity of the Rubisco protein (Figure 2). The plant faced with the impossibility of finding potent sinks to which to direct the photosynthesized, accumulates them in the form of sugars and starch which would inhibit the photosynthesis. In addition, physical obstruction by the starch granules in the chloroplast and the sequestration of phosphorus by the phosphate sugars in the stroma must be taken into account. The second hypothesis explains the lower photosynthetic capacity in response to limitations in the availability of N. In order for the high [CO_2] to stimulate growth, the plant requires a large amount of N that it cannot obtain, causing mismatches in the C/N ratio. Increasing the C/N ratio of these plants would lead to a readjustment of N sinks. Nitrogen would be directed to more limiting processes (photosystems, root growth, etc.) to increase plant sinks.

According to a large number of authors, exposure of nitrogen-fixing plants to elevated [CO_2] will stimulate plant level nitrogen fixation as a result of an increase in the amount of photo-assimilates for nodule formation (Aranjuelo et al., 2008) the enzymes involved in fixation (Cabrerizo et al., 2000). Nodule activity depends on photosynthates supplied by the plant, which are used by the nitrogenase enzyme as a source of energy and reducing power to fix N_2 (Larrainzar et al., 2009). This cycle coupling causes nitrogenase activity in plants to be regulated by photosynthesis (carbon supply), N availability (N source strength) and N demand (N sink strength) (Aranjuelo et al., 2011). Similarly, the products of N_2 fixation are exported throughout the plant via the xylem to other organs where N is required; for example, protein and or osmoregulant synthesis (Ladrera et al., 2007; Aranjuelo et al., 2011). This coupling results in the regulation of nitrogenase activity in plants by photosynthesis (carbon supply), nitrogen availability (N source strength), and N demand (N sink strength). This growth enhancement associated with exposure to elevated CO_2 may lead to a higher N demand. In this context, legumes are particularly interesting due to their symbiotic relationship with N_2-fixing bacteria, which provides them with N autonomy.

Factors favoring the contribution of carbohydrates to legumes have a positive effect on the formation of nodular biomass and number of nodules (Murphy, 1986). The N_2 fixation process in charge of the nodules is an important sink of carbon, energy, etc., for which the plant directs a large number of photoassimilates (Aranjuelo et al., 2014). Taking into account that the high [CO_2] allows a greater availability of photo-assimilates, it is expected that plants grown under high [CO_2] conditions can develop more and larger roots and nodules that allow to fix more N_2. The study conducted by Aranjuelo et al., (2008, 2014), confirmed that soluble sugar and starch concentration of the nodule increased availability of photo-assimilates of the plants grown at 700 $\mu molmol^{-1}$. Previous studies also show that high [CO_2] modifies the nodular metabolism, through the increase in the synthesis, translocation or exudation of organic compounds to the medium favoring the infection and consequently the formation of nodules. N_2 binding will depend on the amount of nodular biomass formed and/or the specific biochemical activity of the nodule. A study conducted in *Medicago sativa* exposed to 700 μmol mol^{-1} [CO_2] revealed that although plant level N_2 fixation increased under elevated [CO_2], the larger photoassimilate availability did not contribute to increases in specific nodule N_2 fixation (Aranjuelo et al., 2008). Moreover, carbohydrate availability decreased in nodules of plants exposed to elevated [CO_2]. According to the same study, the fact that malate dehydrogenase (MDH), phosphoenolpyruvate carboxylase (PEPc) and isocitrate dehydrogenase (ICDH) activities decreased suggests that the deteriorated respiratory mechanism also was involved in the decline in nodule performance. Furthermore, as observed by other studies (Nomura et al., 2006; Fischinger and Schulze, 2010), the lower activity of these enzymes might be linked to the decrease in dicarboxylic acids with key C-skeleton functions, which includes the malate required for N assimilation. These studies suggested that organic acid limitations were mainly attributed to compounds involved in N assimilation.

In addition to affecting legume production, the above mentioned imbalances in the C/N ratio observed in plants grown under high [CO_2] has been described to affect negatively legume quality (Sanz-Saez et al., 2012). Moreover, lower protein content and mineral impoverishment have also been detected in aboveground organs of N_2 fixing alfalfa plants. As previously noted, legumes are important sources for carbohydrates, proteins, amino acids, lipids and minerals. Consequently, they might determine the value of nutritional diet and the quality of animal and human consumption (Sanz-Sáez et al., 2012). Therefore, and taking into account the different climatic scenarios presented by the IPCC (in which the concentration of [CO_2], temperature and periods of drought will increase), it is likely that, in addition to production, the quality of legumes will be negatively affected in the short/medium term future.

4. Perspectives

The expansion of the world population requires the development of agricultural production adapted to present and future environmental conditions and demands. European agricultural production needs to maintain a high level of productivity that is economically and ecologically sustainable. While the increase in fertilizer use has been successful in increasing crop yields in the past, it has not taken into account the environmental risks, often associated with over-fertilization. Therefore, it is necessary to develop new strategies and tools that allow a more efficient management of N in worldwide agriculture. These policies would aim at ensuring high levels of production but by reducing harmful environmental impacts. For this, it is essential to increase our knowledge about the mechanisms that condition symbiotic N_2 fixation and its potentiation as a natural source of nitrogen enrichment of soils, in a context of Climate Change together with the genetic improvement of legumes with increased tolerance of nitrogen fixation to environmental constrains.

References

Alexandratos, N. and J. Bruinsma. 2012. World agriculture towards 2030/2050: the 2012 revision. Rome, FAO.ESA Working paper No. 12-03.

Annicchiarico P., Pecetti L., Abdelguerfi A., Bouizgaren A., Carroni A. M., Hayek T., Bouzina M. M. and Mezni M. 2011. Adaptation of landrace and variety germplasm and selection strategies for lucerne in the Mediterranean basin. *Field Crops Research* 120, 283-291.

Aranjuelo I., Irigoyen J. J. and Sánchez-Díaz M. 2007. Effect of elevated temperature and water availability on CO_2 exchange and nitrogen fixation of nodulated alfalfa plants. *Environmental and Experimental Botany* 59, 99-108.

Aranjuelo I., Irigoyen J. J., Sánchez-Díaz M. and Nogués S. 2008. Carbon partitioning in N_2 fixing *Medicago sativa* plants exposed to different CO_2 and temperature conditions. *Functional Plant Biology* 35, 306-317.

Aranjuelo I., Irigoyen J. J., Nogués S. and Sánchez-Díaz M. 2009. Elevated [CO_2] and water-availability effect on gas exchange and nodule development in N_2-fixing alfalfa plants. *Environmental and Experimental Botany* 65, 18-26.

Aranjuelo I., Molero G., Erice G., Avice J. C. and Nogués S. 2011. Plant physiology and proteomics reveals the leaf response to drought in alfalfa (*Medicago sativa* L.). *Journal of Experimental Botany* 62, 111-123.

Aranjuelo I., Arrese-Igor C. and Molero G. 2014. Nodule performance within a changing environmental context. *Journal of Plant Physiology* 98, 32-39

Cabrerizo P. M., González E. M., Aguilar A., Aparicio-Tejo P. M. and Arrese-Igor C. 2000. Long term effects of CO_2 enrichment on nitrogen fixation and nodule metabolism in *Pisum sativum* epidermal peels correlates with stomatal response to ABA. En: *Nitrogen fixation: From molecules to crop productivity*. pp. 110. Eds. FO Pedrosa, M Hungria, M Geoffrey Yates, WE Newton. Kluwer Academic Publishers. Holland.

Chaves M. M. Pereira J. S., Maroco J., Rodrigues M. L., Ricardo C. P. P. and Osório M. L., et al., 2002. How plants cope with water stress in the field: photosynthesis and growth. *Annals of Botany* 89, 907–16.

Ciais P., Reichstein M., Viovy N., Granier A., Ogee J., Allard V., Aubinet M., Buchmann N., Bernhofer C., Carrara A., Chevallier F., De Noblet N., Friend A. D., Friedlingstein P., Grünwald T., Heinesch B., Keronen P., Knohl A., Krinner G., Loustau D., Manca G., Matteucci G., Miglietta F., Ourcival J. M., Papale D., Pilegaard K., Rambal S, Seufert G., Soussana J. F., Sanz M. J., Schulze E. D., Vesala T. and Valentini R. (2005) Europe-wide reduction in primary productivity caused by the heat and drought in 2003. *Nature* 437, 529–533.

De Luis I., Irigoyen J. J. and Sánchez-Díaz M. 1999. Elevated CO_2 enhances plant growth in droughted N_2-fixing alfalfa without improving water status. *Physiologia Plantarum* 107, 84-89.

Erice G., Sanz-Sáez A., Aranjuelo I., Irigoyen J. J., Aguirreolea J., Avice J. and Sánchez-Díaz M. 2011. Photosynthesis, N_2 fixation and taproot reserves during the cutting regrowth cycle of alfalfa under elevated CO_2 and temperature. *Journal of Plant Physiology* 168, 2007-2014.

Fischinger S. A., Hristozkova M., Mainassara Z. A. and Schulze J. 2010. Elevated CO_2 concentration around alfalfa nodules increases N_2 fixation. *Journal of Experimental Botany* 61, 121–30.

Flexas J., Medrano H. 2002. Photosynthetic responses of C_3 plants to drought. In '*Advancesin Plant Physiology IV*' (Hemantaranjan A Ed.), Scientific Publishers: Jodhpur.

Flexas J., Bota J., Escalona J. M., Sampol B. and Medrano H. 2002. Effects of drought on photosynthesis in grapevines under field conditions: an evaluation of stomatal and mesophyll limitations. *Functional Plant Biology* 29, 461-471.

Fischer R. A., Byerlee D. and Edmeades G. O. 2009. Can Technology Deliver on the Yield Challenge to 2050? Expert Meeting on How to feed the World in 2050. Rome: Food and Agriculture Organization of the United Nations, Economic and Social Development Department.

Graham P. H. and Vance C. P. 2009. Legumes: importance and constraints to greater use. *Plant Physiology* 131: 872-877.

Ladrera R., Marino D., Larrainzar E., González E. M. and Arrese-Igor C. 2007. Reduced carbon availability involved in the nitrogen fixation response to early drought in soybean. *Plant Physiology* 145, 539–46.

Larrainzar E., Wienkoop S., Scherling C., Kempa S, Ladrera R., and Arrese-Igor C. et al., 2009. Carbon metabolism and bacteroid functioning are involved in the regulation of nitrogen fixation in Medicago truncatula under drought and recovery. Molecular Plant Microbe Interaction 22, 1565–76.

Lawlor D. W. and Cornic G. 2002. Photosynthetic carbon assimilation and associated metabolism in relation to water deficits in higher plants. Plant Cell and Environment 25, 275–294.

Lawlor D. W. and Tezara W. 2009. Causes of decreased photosynthetic rate and metabolic capacity in water-deficient leaf cells: a critical evaluation of mechanisms and integration of processes. *Annals of Botany* 103, 543–549.

Long S. P., Ainsworth E. A., Rogers A. and Ort D. R. 2004. Rising atmospheric carbon dioxide: Plants face the future. *Annual Review of Plant Biology*, 55, 591–628.

Medrano H., Escalona J. M., Bota J., Gulías J. and Flexas J. 2002. Regulation of photosynthesis of C_3 plants in response to progressive drought: stomatal conductance as a reference parameter. *Annals of Botany* 89, 895-905.

McDonald, G. K. and Paulsen G. M. 1997. High temperature effects on photosynthesis and water relations of grain legumes. *Plant and Soil* 196, 46-58.

Mulvaney R. L., Khan S. A. and Ellsworth T. R. 2009. Synthetic nitrogen fertilizers deplete soil nitrogen: a global dilemma for sustainable cereal production. *Journal of Environmental Quality* 38, 2295–314.

Murphy P. M. 1986. Effect of light and atmospheric carbon dioxide concentration on nitrogen fixation by herbage legumes. *Plant and Soil* 95: 399-409.

Nomura M., Mai H. T., Fujii M., Hata S., Izui K. and Tajima S. 2006. Phosphoenolpyruvate carboxylase plays a crucial role in limiting nitrogen fixation in *Lotus japonicus* nodules. *Plant and Cell Physiology* 47, 613–21.

Oury F., Godin C., Mailliard A., Chassin A., Gardet O., Giraud A., Heumez E., Morlais J.-Y., Rolland B., Rousset M., Trottet M. and Charmet G. 2012. A study of genetic progress due to selection reveals a negative effect of climate change on bread wheat yield in France. *European Journal of Agronomy* 40, 28–38.

Rawson H. M., Gifford R. M. and Condon B. N. 1995. Portable temperature gradient chambers for global environment change I. Portable chambers for research on short-stature vegetation. *Plant, Cell and Environment* 18, 1048-1054.

Rogers A., Ainsworth E. A. and Leakey A. D. B. 2009. Will elevated carbon dioxide concentration amplify the benefits of nitrogen fixation in legumes? *Plant Physiology* 151, 1009-1016.

Sanz-Sáez A., Erice G., Aguirreolea J., Muñoz F., Sánchez-Díaz M., and Irigoyen J. J. 2012. Alfalfa forage digestibility, quality and yield under future climate change scenarios vary with *Sinorhizobium meliloti* strain. *Journal of Plant Physiology*, 169: 782-788.

Schulze J. 2004. How are nitrogen fixation rates regulated in legumes. *Journal of Plant Nutrition and Soil Sciences* 167: 125-137.

Serraj R., Sinclair T. R. and Purcell L. C. 1999. Symbiotic N_2 fixation response to drought. *Journal of Experimental Botany* 50, 143-155.

Zahran H. H. 1999. Rhizobium-legume symbiosis and nitrogen fixation under severe conditions and in an arid climate. *Microbiology and Molecular Biology Reviews* 63, 968-989.

In: Legumes for Global Food Security ISBN: 978-1-53612-265-7
Editors: Alfonso Clemente et al. © 2017 Nova Science Publishers, Inc.

Chapter 3

Diversity and Conservation of Wild and Primitive Common Bean Germplasm and Their Associated Rhizobia in the Andean Region

Antonio M. De Ron[1,*], Mariana J. Ferreyra[2], M. Carmen Menéndez-Sevillano[2], Luis Ibarra[2], Gustavo González Anta[3], Diego Perrig[4], A. Paula Rodiño[1] and Eulogio J. Bedmar[5]

[1]Biology of Agrosystems, Misión Biológica de Galicia (MBG), Spanish National Research Council (CSIC), UA Sistemas Agroforestales, Pontevedra, Spain

[2]Active Bank of Northwestern Argentina (BANOA), EEA Salta, National Agricultural Technology Institute (INTA), Salta, Argentina

[3]Rizobacter, Pergamino, Argentina

[4]Indrasa Biotecnología, Río Cuarto, Argentina

[5]Department of Microbiology, Estación Experimental del Zaidín (EEZ), Spanish National Research Council (CSIC), Granada, Spain

[*] Corresponding Author: amderon@mbg.csic.es.

ABSTRACT

Conservation of the genetic resources for agricultural development and global future food and feed security is imperative, including cultivated species, wild relatives and associated microbiota. Common bean (*Phaseolus vulgaris* L.) is the most important grain legume for direct human consumption on a global scale and this crop shows great phenotypic variation, which enables its production in a wide range of agrosystems and cropping systems. In the Northwest of Argentina (NOA) there is still a considerable amount of genetic diversity in many wild populations and also in the traditional varieties or primitive landraces and many of them are maintained at the Active Bank of Northwestern Argentina (BANOA). This germplasm is studied in different projects and some actions are being taken to preserve them *in situ* and *ex situ*, as presented in this work. The common bean is considered to be a poor nitrogen fixing plant when compared to other grain legumes. Although so far *R. etli* symbiovar *phaseoli* is the predominant species in root nodules of wild and cultivated common beans of Andean origin, investigations of the bean genetic background involved in nodulation and nitrogen fixation efficiency might result in clues for applied aspects of the *P. vulgaris*-rhizobia symbiosis. Among them, search for rhizobial (and non-rhizobial) strains with better performance to increase biological N_2-fixation under specific environmental conditions is mandatory. Mixed cropping of beans and maize is a common practice in the Andean area. Regarding the quality of the bean grain, especially in its protein content as grain legume, there are also certain aspects that are not definitive to affirm that it is superior in intercropping and with regard to efficiency of the symbiotic system with rhizobia, and the fixation and use of nitrogen; it seems that intercropping is more beneficial for both crops. Thus, the association of crops can help extend the cultivation of bean and maize, especially in family agriculture, which is of great relevance in many countries as a secure and sustainable food source of healthy protein. In conclusion, it can be considered that in the Andean zone, especially in the NOA, there is an interesting diversity both in the wild populations and in the primitive or domesticated varieties of common bean, together with the associated microbiota, and this germplasm, object of study in several projects, deserves to be conserved, *in situ* and *ex situ*, both as a genetic heritage and for its potential value for genetic improvement.

Keywords: biodiversity, genetics, germplasm, intercropping, plant breeding, *Phaseolus vulgaris* L, *Rhizobium,* symbiosis

1. Introduction

Genetic resources for agricultural development and global future food and feed security are constantly threatened; therefore their conservation is imperative. *In situ* conservation of genetic resources ensures continuity of the dynamic processes in which species develop under environmental changes and co-evolve with abiotic and biotic factors, such as their associated microbiota, so the processes of evolution and adaptation are maintained and the generation of new genotypes is guaranteed as well as the genetic integrity of the cultivated varieties of crops and their wild relatives. On-farm conservation is a type of *in situ* germplasm conservation, being farmers and local communities the responsible for maintaining the genetic diversity in agrosystems where they preserve their traditional varieties of crops and the ancestral knowledge associated with them. By the other hand, *ex situ* conservation of cultivated plants, their wild relatives and the associated microbiota in genebanks ensures the long-term conservation of them, although relevant changes in the genetic material should be avoided.

Common bean (*Phaseolus vulgaris* L.) is the most important grain legume for direct human consumption on a global scale (Day, 2013). This crop shows great variation in phenology and plant and seed phenotype, which enables its production in a wide range of agrosystems and cropping systems in The Americas, Africa, Middle East, China and Europe. Beans are consumed mainly as a grain legume, due to the high protein content of the grain, but the use of the fresh pods and seeds as vegetables is frequent in many countries. Common bean is the main source of plant protein in many areas of Africa and Latin America, being also a relevant component in traditional diets in the Middle East and the European Mediterranean region as a component of healthy diets (Broughton et al. 2003, Casquero et al. 2006, De Ron et al. 2015). In addition, it is gaining importance in USA where bean consumption has increased due to growing interest in "ethnic" and healthy food (Blair and Izquierdo, 2012).

The recent developments in the knowledge of the symbiotic interaction between different varieties of common bean and different rhizobia strains will contribute to the more efficient use of this crop and its associated microbiota in order to achieve safer and more sustainable food production. By the other hand, the integration of the -omic data currently available into the genebanks documentation systems and its combination with previous genetic and phenotypic data will open the future for the valorization of the common bean genetic resources and their use as a healthy future food source (Schmutz et al. 2014, Vlasova et al. 2016).

2. Origin, Domestication and Evolution of the Common Bean

The bean crop expanded from The Americas to all the continents for centuries resulting in a complex genetic structure of the bean germplasm outside its areas of origin and domestication (South and Central America) (De Ron et al. 2015). The diffusion of the common bean out of the American domestication centers was a complex process involving introductions into different continents and countries and some of them have been proposed as secondary centers of diversification, such as Europe (Santalla et al. 2002, Angioi et al. 2010, Gioia et al. 2013), Central-eastern and Southern Africa, Brazil and China (Bellucci et al. 2014).

Andean domesticated common bean germplasm displays high diversity in plant and seed morphology and in adaptation to different agrosystems, but it has been proved to have a narrow genetic base according to molecular analyses as compared to Mesoamerican germplasm. The genetic structure of common bean landraces reflects the structure of the wild bean populations, which means that the narrow genetic variability in the Andean domesticated germplasm is also found in the wild germplasm (Koenig and Gepts 1989, Cattan-Toupance et al. 1998).

Understanding the effects of domestication on genetic diversity of common bean is of great importance, not only for the knowledge of the crop evolution process but also for possible applications, such as the implementation of appropriate biodiversity conservation strategies and the use of genetic variability in breeding programs. One of the most important and generalized features of plant domestication is the reduction in genetic diversity, not only during the initial domestication process but also during dispersion and adaptive radiation from the centers of domestication to other areas. The reduction of genetic diversity is usually more drastic in selfing species such as the common bean, which have limited genetic recombination, compared with the open pollinated species (Jarvis and Hodgkin 1999). In many crops, this reduction is due to stochastic events (bottlenecks and genetic drift due to reduction of population size) and selection (adaptation to novel agrosystems) (Vigouroux et al. 2002).

The geographical area of domestication for the Andean common bean gene pool as well as the number of domestication events has been controversial although previous studies have indicated that Bolivia is the main candidate for the primary domestication site of the Andean gene pool of common bean (Beebe et al. 2001). Argentinean domesticated germplasm includes mostly bean types of the races Nueva Granada and Peru; however, races Chile and Mesoamerica have also been found (Singh et al. 1991) which could be due to germplasm exchanges in pre-Columbian times between Mesoamerica and South-America although extensive seed movement probably occurred after the 1500s.

During its domestication the common bean evolved from small seeded types with indeterminate climbing growth habit to medium-large seeded forms with determinate growth habit. The domestication affected also to the seed dormancy since the seed coats of domesticated or cultivated varieties are permeable to water, therefore, no dormancy is present in them. Finally, the dehiscence of the fruit based upon high fibrous walls is not present in the domesticated forms (Brücher and Brücher 1976, Smartt 1988, Gepts and Debouck 1991). The reduction of the content in some toxic materials in the domestication process benefits the seed flavor but caused the loss of seed protection against predators (Smartt 1988). As in other crops, the founder effect (genetic drift) and the unconscious selection by farmers in many *loci* resulted in a reduction in the genetic diversity in the domesticated bean varieties (De Ron et al. 2015).

Figure 1. Seeds of wild (left), primitive (center) and improved (right) common bean varieties.

The clear phenotypic differences among the Andean bean germplasm (Figure 1), due to the domestication syndrome (Hammer 1984, Koinange et al. 1996), resulted from selection during many years for adaptation to different agrosystems, together with the already mentioned genetic drift. Currently, the wild bean Andean populations display more genetic variability than the domesticated varieties (Gepts et al. 1986, Koenig et al. 1990). This fact suggests a strong founder effect in the domestication process affecting the wild populations (Ladizinsky 1985), which could have eroded valuable genetic variability of wild beans not present currently in the domesticated varieties (van Schoonhoven et al. 1983, Debouck and Tohme 1989).

The relevance of wild bean populations for the genetic improvement of cultivated varieties is based upon the fact that the progeny of crosses between wild and domesticated bean plants is fully fertile and there are no major barriers for these crosses (Cattan-Toupance et al. 1998, Singh et al. 1995, Singh 2001). An example of the potential of wild types as a source of genetic diversity for breeding is the case of the arcelin seed protein that protects the seeds against the Mexican bean weevil (*Zabrotes subfasciatus* Boheman), being this character absent in many of the cultivated varieties of the Andean common bean (Schoonhoven and Cardona 1982), and only some wild bean Mesoamerican bean populations from Mexico are highly resistant to that pest (Acosta-Gallegos et al. 1998). For these reasons, the study of the diversity of Andean bean germplasm including wild and domesticated types could display valuable genetic variation to broaden the genetic base of Andean common bean cultivated germplasm that will benefit improvements in pest and diseases resistance and productivity, frequently hard to achieve by selection in the cultivated varieties (White et al. 1992, Kornegay et al. 1992).

Regarding the history of the evolution of beans in the Andean area, a special case is the nuña bean, considered as an early-domesticated common bean type, traditionally grown in the Andean highlands in pre-ceramic times (Tohme et al. 1995). Observations of ancient beans at the Guitarrero Cave in Ancash, Peru, indicate that nuña beans may have been available 11,000 years ago (Kaplan and Kaplan 1988). Its seeds may have been the first bean consumed by man since they can easily be prepared for eating by roasting or toasting them on a heated stone (Figure 2). The first selection pressures leading to domestication of common bean could have resulted in the development of nuña beans, and maybe that toasting grains was a well-established tradition in the Andes and possibly in Mesoamerica, where early maize races have also been used for popping (Mangelsdorf and Smith 1949). However, no evidence of nuña beans has been found in Mesoamerica, probably due to genetic differences in the photoperiod response between the Mesoamerican and Andean gene pools (Kornegay et al. 1993, Tohme et al. 1995). Nuña bean could be considered as the bean counterpart of popcorn since both are ancient forms of their respective crops, usually cultivated in intercropping in the Andean region, and their grains expand or roast, in different ways, by heat. The reason for the process of expansion by heat is not well-known, but it is currently under research (González et al. 2014).

Figure 2. Crude (left) and roasted (right) nuña beans.

3. Diversity and Conservation of Wild and Primitive Andean Common Bean Germplasm

Argentinian domesticated bean varieties were evaluated by De la Fuente et al. (2013) who found a relevant genetic variation on them with a considerable number of unique haplotypes and private alleles and this germplasm seems to integrate into two homogeneous and distinct groups (Tohme et al. 1996, De Ron et al. 2004, Santalla et al. 2004, Galván et al. 2006). In their study, De la Fuente et al. (2013) also observed a clear reduction in genetic variation in wild and domesticated accessions in Argentinean germplasm with many of the alleles of the wild types shared with the domesticated ones and some loci showing more variation in domesticated varieties than in wild populations.

Some qualitative phenotypic traits were homogeneous in the wild populations and in the primitive landraces: type IV indeterminate climbing growth habit (CIAT 1984), colored flowers and colored seeds. The mottled seed coat pattern in Argentinean wild bean seeds is considered as a phylogenetically ancient feature with a marked selection advantage against animal predators such as birds and rats when they fall onto a debris-covered soil.

Figure 3. Provinces of the Northwestern Argentina (NOA).

In the Northwest of Argentina (NOA, Figure 3) there is still a considerable amount of genetic diversity in many wild populations and also in the traditional varieties or primitive landraces, owing to the small size of the farms in these areas and the persistence of traditional methods of agriculture. Domesticated common bean populations in the NOA, a complex germplasm displaying wide variation in phenotypic traits (Menéndez-Sevillano 2002, De Ron et al. 2004), maybe as a consequence of the evolution of the species after domestication and its coexistence with wild ancestral forms. Debouck (1989) proposed that the phenotypic variation in the bean landraces seeds in the NOA could be originated by "aesthetic" selection for non-food uses of the dry seeds in this area. This aesthetic selection would have lasted for several millennia resulting in types with a wide variation in seed colors and color patterns. Burkart (1955) reported this aesthetic use of beans in Bolivia as a children toy called "chuis" due to the vividly colored seeds.

Different collecting missions involving common bean primitive landraces and wild populations were carried out in the NOA since the 70's (Menéndez-Sevillano 2002) in order to increase the availability of regional genetic diversity that could be used as a source of genes to introduce in modern varieties (Singh et al. 1995). The Active Bank of Northwestern Argentina (BANOA) is one of the genebanks of the INTA Genetic Resources Network where the common bean collection, which includes 221 wild populations and 401 primitive landraces, is the main germplasm collection of that genebank. The knowledge of the phenotypic and genotypic variation of the wild common bean germplasm is a prerequisite for their *in situ* conservation that means to preserve as far as possible the genetic integrity of the wild populations as a relevant natural resource in the areas where they grow usually.

To know the climatic characteristics of the locations where the wild populations and the primitive landraces grow helps to define the choice of germplasm accessions by breeders. In the context of bean production in Argentina, where this crop is being displaced by soybean to regions of higher temperature and lower precipitation, this information is of great value.

The most variable wild bean populations were found in the Yungas Forest Ecoregion of Argentina (Los Toldos, Cuesta del Clavillo and Higuerillas); therefore, these populations would be the most appropriate to establish genetic reserve areas for wild relatives of cultivated beans. Dry Chaco populations located in the isohyets 600-800 mm rainfall would be suitable for incorporation in bean breeding programs aimed to finding water stress tolerance.

With regard to domesticated or primitive landraces, participative maintenance by reinsertion of these bean landraces into farmer fields under traditional growing methods (such as intercropping with maize) is an initiative in the NOA that will contribute to the conservation of this ancient germplasm. Complementarity between *ex situ* and *in situ* conservation will clearly result in greater efficiency in maintaining local bean varieties. To enhance the participation of farmers, actions to achieve greater productivity of their varieties are needed on the basis of their knowledge, using local technology and strong personal involvement of such farmers. Articulation between the germplasm bank and the regional agricultural technicians of the NOA includes activities such as: i) delivering of bean seeds from BANOA collection to be locally distributed, ii) visits to farmers growing the distributed varieties, iii) local training of farmers about the maintenance of the genetic resources and the relevance of preserve their own seeds, and iv) contribution to local seed fairs.

4. Diversity of Symbiotic Rhizobia Nodulating *Phaseolus Vulgaris*

Within the angiosperms, legumes (family Fabaceae or Leguminosae) are well-known by their ability to establish dinitrogen (N_2)-fixing symbioses with soil bacteria collectively referred to as rhizobia. As a consequence of the plant-bacteria interaction, a new organ, the nodule (Figure 4), is formed in the roots, rarely in the shoots and leaves, where N_2 fixation takes place. After induction by flavonoids produced by host legumes, rhizobia synthesize the so called nodulation factor (Nod) which is excreted to the surrounding medium where it induces the plant signal transduction pathway that results in formation of infection threads development and nodule morphogenesis. Nod factors that are recognized by the different host plants vary between rhizobial species and are a basis for host-symbiont specificity. The exchange of molecular signals that occurs between the two partners leads finally to the formation of the nodules, where nitrogen fixation takes place (Gough and Cullimore 2011; Oldroyd 2013).

Pure bacterial cells from legume nodules capable of reducing atmospheric N_2 were first obtained by Beijerink in 1888, and one year later Frank named them *Rhizobium leguminosarum*. The name rhizobia is used since then to mean Gram negative aerobic rods with the capacity to induce nodules, where they transform into specialized cells called bacteroids, and fix N_2. The bacteroids synthetize the enzyme nitrogenase which is responsible for the reduction of molecular nitrogen (N_2) to ammonia (NH_4^+) according to the following reaction [1]:

$$N_2 + 8\,e^- + 8\,H^+ + 16\,ATP \rightarrow 2NH_4^+ + H_2 + 16\,ADP + 16\,Pi \qquad [1]$$

Figure 4. Nodules in common bean roots.

Nodulation is common in the subfamily Faboideae (or Papilionoideae), less abundant in subfamily Mimosoideae and seldom in Cesalpinoideae (Sprent, 2007; Sprent et al. 2017). Ammonia formed during activity of nitrogenase is used to form amino acids, proteins, hormones and other nitrogenous compounds required for plant growth and development. This is why legumes, in association with their rhizobia, can grow in arid, unfertile soils with low N content.

Morphological, physiological and symbiotic characteristics have been used for definition of rhizobial species. As occurred in all groups of bacteria, the species concept in rhizobia is linked to the advent of genomic techniques and drastically changed after the proposal of using the ribosomal 16S (*rrs*) (16S rRNA) gene, which encodes the synthesis of the small 16S subunit of the ribosomes, as a universal tool for the identification and classification of bacteria. Together with the *rrs* gene, knowledge of a number of genes involved in different cell functions has permitted selection of core genes for species differentiation. The housekeeping genes *atpD*, which encodes the beta subunit of the membrane ATP synthase that is essential for energy production, *recA*, which is involved in DNA recombinational repair, *glnII*, responsible for synthesis of the enzyme that catalyzes the ATP-dependent biosynthesis of glutamine from glutamate and ammonia, *rpoB*, which encodes the β subunit of bacterial RNA polymerase, among others, were originally proposed by Gaunt et al. (2001) and are now widely used.

Based on phenotypic and molecular analyses all known rhizobia isolated from root nodules until the end of the XX century were members of the order Rhizobiales of the Alphaproteobacteria. Nodulation of legumes by non-rhizobial Alphaproteobacteria species was shown for the first time in 2001 and clearly demonstrated afterwards (see reviews by Graham 2008; Rivas et al. 2009, Velazquez et al. 2010; Peix et al. 2015; Shamselding et al. 2017). Also in 2001, *Burkholderia* and *Cupriavidus* species of the Betaproteobacteria were isolated by Moulin et al. (2001) and Chen et al. (2001), respectively, thus showing that legume nodulation was not restricted to Alphaproteobacteria. Gammaproteobacteria of the genera *Pantoea*, *Enterobacter*, *Escherichia*, *Leclercia* and *Pseudomonas* have also been isolated from root nodules of several legumes, but their presence and role is yet to be defined.

Whereas some legumes such as *Cicer*, *Vicia* and *Trifolium* are restrictive hosts being nodulated only by a specific rhizobia, others are promiscuous legumes that can be nodulated by several species of rhizobia such as *Phaseolus* (Michiels et al. 1998). There is to note, however, that this promiscuity concept should not be based on the number of taxonomic rhizobial species able to nodulate a legume, but on different symbiotic genes able to induce the nodulation process (Nod factors). Like the legumes, rhizobia can have a broad or a narrow host range as some strains can nodulate only legumes of a given genus whereas others nodulate more than one host plant. At the time of the writing of this article, 21 rhizobial species have been isolated from nodules of *P. vulgaris*, including species of genera *Rhizobium*, *Ensifer* (formerly *Sinorhizobium)*, *Burkholderia* and *Cupriavidus* spanning the Alphaproteobacteria and Betaproteobacteria classes (Table 1) (Velazquez et al. 2010; Peix et al. 2015; Shamselding et al. 2017).

The term symbiovar groups strains within a species that are able to establish symbiosis with a specific legume. This concept is currently defined based on the range of legumes a symbiovar can nodulate, and mainly on the basis of the *nodC* gene analysis, whose phylogenies make possible the correct differentiation of these categories within species of all bacteria-nodulating genera. Being *P. vulgaris* a promiscuous legume, bacterial strains isolated from its nodules harbor divergent *nodC* gene and it belongs to different symbiovars (Peix et al. 2015 and references therein).

R. etli has been considered for decades the predominant species found associated with both wild and cultivated *P. vulgaris* in the distribution centers of common bean in America (Segovia et al. 1993; Bernal and Graham 2001; Silva et al. 2003); the current available data, however, suggest that *R. phaseoli* is also an abundant endosymbiont of common bean in American countries and that both *R. etli* and *R. phaseoli* species have a common American origin. Accordingly, it has been proposed that those species coevolved with *P. vulgaris* in its diversification centers and were disseminated in other geographical sites with common bean seeds (Díaz-Alcántara et al. 2014).

Table 1. Endosymbiotic bacteria isolated from root nodules of *P. vulgaris*

Class α-Proteobacteria	References
Genus: *Rhizobium*	
Species	
R. acidisoli	Román-Ponce et al. 2016
R. azibense	Mnasri et al. 2014
R. etli symbiovar phaseoli	Segovia et al. 1993
R. freirei	Dall'agnol et al. 2013
R. gallicum symbiovar phaseoli	Amarger et al. 1997
R. giardinii symbiovar phaseoli	Amarger et al. 1997
R. leguminosarum symbiovar phaseoli	Jordan 1984; García-Fraile 2010
R. leucaenae	Ribeiro et al. 2012
R. lusitanum	Valverde et al. 2006
R. mesomaericanum	López-López et al. 2012
R. paranaense	Dall'agnol et al. 2014
R. phaseoli	Dangeard 1926; Ramírez-Bahena et al. 2008
R. tropici	Martínez-Romero et al. 1991
R. vallis	Wang et al. 2011
Genus: *Ensifer*	
E. americanum symbiovar mediterranense	Mnasri et al. 2007
E. fredii symbiovar mediterranense	Mnasri et al. 2012
E. meliloti symbiovar mediterranense	Mnasri et al. 2007
Class: β-Proteobacteria	
Genus: *Burkholderia*	
B. caballeronis	Martínez-Aguilar et al. 2013
B. phymatum	Vandamme et al. 2002; Elliott et al. 2007; Talbi et al. 2010; Bournaud et al. 2013
B. tuberum	Vandamme et al. 2002; Elliott et al. 2007
Cupriavidus necator	da Silva et al. 2012

Bacterial strains isolated from nodules of *P. vulgaris* all over the world carry two divergent *nodC* genes involved in nodulation that have been named α and γ alleles, and different works have established they are of Mesoamerican and Andean origin, respectively (Ribeiro et al. 2013; Díaz-Alcántara et al. 2014). Species *R. etli* and *R. phaseoli* contain strains harboring the *nodC* gene γ allele, and for those strains the name symbiovar phaseoli has been proposed.

The presence of rhizobial cells, specifically *R. etli*, in the testa of bean seeds has been reported (Pérez-Ramírez et al. 1998), so that rhizobia could have been spread with *P. vulgaris* seeds in Europe and from here to Africa. This assumption was clearly established after the finding that the γ allele of the *nodC* gene was present in nodules of species *R. leguminosarum*, *R. lusitanum*, *R. gallicum* and *R. giardinii* in Portugal, Austria, France and Spain (Sessitsch et al. 1997; Amarger et al. 1997; Valverde et al. 2006, García-Fraile et al. 2010; Mulas et al. 2011). These data agree with the scenario that mostly Andean cultivar of *P. vulgaris* have been traditionally cultivated in Europe (Díaz-Alcántara et al. 2014).

In addition to rhizobia, a variable microbiome can be found in legume nodules where they co-inhabit and play a usually unknown ecological role with respect to the plant (Peix et al. 2015). Information is lacking on the role of legume nodular endophytes, both during the infection process by rhizobia and their possible contribution to plant growth. Interestingly, rhizobial cells can also be endophytic in legume nodules, this is, those rhizobia were obtained from nodules of common bean, but were unable to re-infect the legume from which they were isolated. This is the case with *Phyllobacterium endophyticum*, *Cohnella phaseoli*, *Fontibacillus phaseoli* and *Herbaspirillum lusitanum* which were recovered from nodules of *P. vulgaris* but did not re-infect their host plant (Peix et al. 2015, and references therein).

5. The Intercropping Bean-Maize and Its Effect in the Symbiotic System Bean Rhizobia

Mixed cropping of two or more species is the most common form of production in areas where subsistence agriculture is the norm. Such systems are complex and frequently include legumes and cereals (Hamblin and Zimmermann, 1986). In particular, the intercropping system of common bean and maize has been a common practice in Mesoamerica and South America, as well as in the south of Europe (Spain and Portugal) (Santalla et al. 1994).

Santalla et al. (1995, 1999) analyzed the grain quality of bean varieties intercropped with field maize, and in their study, significant variation for most of the quality traits was revealed among common bean varieties. There was no evidence for any differences in relative performance of bean populations in different cropping systems although there were differences between cropping systems for protein content.

The difference between cropping systems for protein content of the bean grain could be interpreted in terms of the ability of maize to enhance the nitrogen accumulation by the bean crop. Clark and Francis (1985) found that nitrogen accumulation in beans in intercrops exceeded that in a sole crop by 30% and that may be due to more efficient nitrogen mobilization when bean varieties grow associated with maize. Nodulation in legumes has been shown to increase when growing with species unable to fix nitrogen, as reported by Boucher and Espinosa (1982) who noticed a greater stimulus for nodulation in beans when intercropped with maize. Hence, maize is more successful at extracting labile nutrients like nitrogen and water from the soil, and furthermore the competition from intercropped maize for nitrogen may stimulate fixation in the common bean (Davis and Woolley 1993, Stem 1993). This may be the fundamental difference with regard to common bean population selection for crude protein. Growth, nodulation and grain yield were assessed by Latati et al. (2016) and correlated with the rhizosphere soil phosphorus availability. Their results showed that phosphorus (P) availability significantly increased in the rhizosphere of both species, especially in intercropping under the P-deficient soil conditions. This increase was associated with high efficiency in use of the rhizobial symbiosis (high correlation between plant biomass and nodulation), plant growth and resource (nitrogen and phosphorus) use efficiency.

A significant interaction of bean genotype x cropping system was found for the number of nodules per plant and nodule moisture on different bean varieties studied, days to emergence, days to flowering, end of flowering, shot length, root dry weight and shoot nitrogen content (Santalla et al. 2001). Nodulation parameters were correlated positively with shoot and root yield and duration of flowering and correlated negatively with grain protein content, pod and seed dimensions and seed dry weight. These observations indicate that it may be possible to increase both the symbiotic nitrogen (N_2)-fixation and the seed yield through plant breeding in intercropping.

Conclusion

As a result of the collections carried out in the NOA, an important number of wild and primitive accessions forms have been collected. Of the total collection conserved in the BANOA, 80% has been characterized morphoagronomically and 25% molecularly and so the information of the variability contained in this collection is available. Wild populations have been identified for *in situ* conservation, having studied in order to know those with greater variability. *In situ* conservation of local varieties has also begun in farmer fields. Although the number of accessions is important, there are still places to prospect, because new characters of interest could be found in germplasm not collected yet.

Despite promiscuity of common bean to establish N_2-fixing symbioses with different rhizobial and non-rhizobial species, *P. vulgaris* is considered to be a poor nitrogen fixing plant when compared to other grain legumes (Ramos and Boddey 1987; Hardarson 1993), a problem that has been attributed to the ineffectiveness of indigenous rhizobia or to environmental constraints. Although so far *R. etli* symbiovar *phaseoli* is the predominant species in root nodules of wild and cultivated common beans of Andean origin, investigations of the bean genetic background involved in nodulation and nitrogen fixation efficiency might results in clues for applied aspects of the *P.-vulgaris*-rhizobia symbiosis. Among them, search for rhizobial (and non-rhizobial) strains with better performance to increase biological N_2-fixation under specific environmental conditions is mandatory.

Regarding the quality of the bean grain, especially in its protein content as grain legume, there are also certain aspects that are not definitive to affirm that it is superior in intercropping. With regard to efficiency of the symbiotic system with rhizobia, and the fixation and use of nitrogen, it seems that intercropping is more beneficial for both crops. Thus, the association of crops can help extend the cultivation of bean and maize, especially in family agriculture, which is of great relevance in many countries as a food source of healthy protein.

In conclusion, it can be considered that in the Andean zone, especially in the NOA, there is an interesting diversity both in the wild populations and in the primitive or domesticated varieties of common bean. This germplasm, object of study in several projects, deserves to be conserved, *in situ* and *ex situ*, both as a genetic heritage and for its potential value for genetic improvement.

ACKNOWLEDGMENTS

The authors thank the financial support from the i-COOP project 2016SU0004 (CSIC, Spain) and also from the National Institute of Agricultural Technology (INTA, Argentina) for the different research projects awarded.

References

Acosta-Gallegos, J. A., Quintero, C., Vargas, J. et al. (1998). A new variant of arcelin in wild common bean, *Phaseolus vulgaris* L., from Southern Mexico. *Genetic Resources and Crop Evolution*, 45: 235-242.

Amarger, N., Macheret, V. and Laguerre, G. (1997). *Rhizobium gallicum* sp. nov. and *Rhizobium giardinii* sp. nov., from *Phaseolus vulgaris* nodules. *International Journal of Systematic and Evolutionary Bacteriology*, 47: 996-1006.

Angioi, S. A., Rau, D., Attene, G. et al. (2010). Beans in Europe: origin and structure of the European landraces of *Phaseolus vulgaris* L. *Theoretical and Applied Genetics*, 121: 829-843.

Beebe, S., Rengifo, J., Gaitan, E., Duque, M. C. and Tohme, J. (2001). Diversity and origin of Andean landraces of common bean. *Crop Science*. 41: 854-862.

Bellucci, E., Bitocchi, E., Rau, D. et al. (2014). Genomics of origin, domestication and evolution of *Phaseolus vulgaris,* In: Tuberosa R, Graner A, Frison E (eds.) Genomics of plant genetic resources. Springer, Dordrecht, pp 483-507.

Bernal, G. and Graham, P.H. (2001) Diversity in the rhizobia associated with *Phaseolus vulgaris* L. in Ecuador, and comparisons with Mexican bean rhizobia. *Canadian Journal of Microbiology*, 47: 526-534.

Blair, M. W. and Izquierdo, P. (2012). Use of the advanced backcross-QTL method to transfer seed mineral accumulation nutrition traits from wild to Andean cultivated common beans. *Theoretical and Applied Genetics*, 125: 1015-103.

Boucher, D. and Espinosa, J. (1982). Cropping systems and growth and nodulation responses of beans to nitrogen in Tabasco, Mexico. *Tropical Agriculture*, 59: 279-282.

Bournaud, C., de Faria, S. M., dos Santos, J. M., Tisseyre, P., Silva, M., Chaintreuil, C., Gross, E., James, E. K., Prin, Y. and Moulin, L. (2013). *Burkholderia* species are the most common and preferred nodulating symbionts of the *Piptadenia* group (tribe Mimoseae). *PLoS One*, 8: e63478.

Broughton, W. J., Hernández, G., Blair, M. et al. (2003). Beans (*Phaseolus* spp.)-model food legumes. *Plant Soil* 252: 55-128.

Brücher, B. and Brücher, H. (1976). The South American wild bean (*Phaseolus aborigineus* Burk.) as ancestor of the common bean. *Economic Botany*, 30: 257-272.

Burkart, A. (1955). Presentación de porotos de adorno de Bolivia. *Physis*, XX 55: 56. [Presentation of ornamental beans from Bolivia. *Physis*, XX 55: 56].

Casquero, P. A., Lema, M., Santalla, M. and De Ron, A. M. (2006). Performance of common bean landraces from Spain in the Atlantic and Mediterranean environments. *Genetic Resources and Crop Evolution*, 53: 1021-1032.

Cattan-Toupance, I., Michalakis, Y. and Neema, C. (1998). Genetic structure of wild bean populations in their South-Andean centre of origin. *Theoretical and Applied Genetics*, 96: 844-851.

Chen, W. M., Laevens, S., Lee, T. M., Coenye, T., de Vos, P., Mergeay, M. and Vandamme, P. (2001). *Ralstonia taiwanensis* spp. nov. isolated from root nodules of *Mimosa* species and sputum of a cystic fibrosis patient. *International Journal of Systematic and Evolutionary Bacteriology*, 51: 1729-1735.

CIAT (1984). Morfología de la planta de fríjol común (*Phaseolus vulgaris* L.). Guía de estudio. CIAT (Centro Internacional de Agricultura Tropical). Cali, Colombia. [Plant morphology of the common bean (*Phaseolus vulgaris* L.). Study guide. CIAT (International Center for Tropical Agriculture). Cali, Colombia].

Clark, A. E. and Francis, C. A. (1985). Transgressive yielding in bean: maize intercrops; interference in time and space. *Field Crops Research* 11: 37-53.

da Silva, K., Florentino, L. A., Barroso da Silva, K. B., de Brandt, E., Vandamme, P. and de Souza-Moreira, F. M. (2012). *Cupriavidus necator* isolates are able to fix nitrogen in symbiosis with different legume species. *Systematic and Applied Microbiology*, 35: 175-182.

Dall'agnol, R. F., Ribeiro, R. A., Ormeño-Orrillo, E., Rogel, M. A., Delamuta, J. R., Andrade, D. S., Martínez-Romero, E., and Hungria, M. (2013). *Rhizobium freirei*, a symbiont of *Phaseolus vulgaris* very effective in fixing nitrogen. *International Journal of Systematic and. Evolutionary Microbiology*, 63: 4167–4173.

Dall'agnol, R. F., Ribeiro, R. A., Delamuta, J. R. M., Ormeño-Orrillo, E., Rogel, M. A., Andrade, D. S., Martínez-Romero, E. and Hungria, M. A. (2014). *Rhizobium paranaense* sp. nov., an effective N_2-fixing symbiont of common bean (*Phaseolus vulgaris* L.) with broad geographical distribution in Brazil. *International Journal of Systematic and. Evolutionary Microbiology*, 64: 3222-3229.

Dangeard, P. A. C. (1926). Recherches sur les tubercles radicaux des L′egumineuses. 1st ed., Du Botaniste, Paris, France. [Research on the radical tubercles of legume plants. 1st ed., Du Botaniste, Paris, France].

Davis, J. H. C. and Woolley, J. N. (1993). Genotypic requirement for intercropping. *Field Crops Reserarch*, 34: 407-430.

Day, L. (2013). Proteins from land plants-Potential resources for human nutrition and food security. *Trends in Food Science and Technology*, 32: 25-42.

De la Fuente, M., González, A. M., De Ron, A. M. et al. (2013). Patterns of genetic diversity in the Andean gene pool of common bean reveal a candidate domestication gene. *Molecular Breeding*, 31:501-516.

De Ron, A. M., Menéndez-Sevillano, M. C. and Santalla, M. (2004). Variation in primitive landraces of common bean (*Phaseolus vulgaris* L.) from Argentina. *Genetics Resources and Crop Evolution*, 51: 883-894.

De Ron, A.M., Papa, R., Bitocchi, E. et al. (2015). Common bean. In: De Ron, A.M. (ed.) Grain legumes, Series: Handbook of Plant Breeding, Springer Science+Business Media, New York, pp 1-36.

Debouck, D. G. (1989). Early beans (*Phaseolus vulgaris* L. and *P. lunatus* L.) domesticated for their aesthetic value?. *Annual Report of the Bean Improvement Cooperative*, 32: 62-63.

Debouck, D. G. and Tohme, J. (1989). Implications for bean breeders of studies on the origins of common bean, *Phaseolus vulgaris* L. In: Beebe S (ed.) Current Topics in Breeding of Common Bean, Working Document 47, CIAT (Centro Internacional de Agricultura Tropical), Cali, pp 3-42.

Díaz-Alcántara, C. A., Ramírez-Bahena, M. H., Mulas, D., García-Fraile, P., Gómez-Moriano, A., Peix, A., Velázquez, E. and González-Andrés, F. (2014). Analysis of rhizobial strains nodulating *Phaseolus vulgaris* from Hispaniola Island, a geographic bridge between Meso and South America and the first historical link with Europe. *Systematic and Applied Microbiology*, 37: 149-156.

Elliott, G. N., Chen, W. M., Chou, J. H., Wang, H. C., Sheu, S. Y., Perin, L., Reis, V. M., Moulin, L., Simon, M. F., Bontemps, C., Sutherland, J. M., Bessi, R., de Faria, S. M., Trinick, M. J., Prescott, A. R., Sprent, J. I. and James, E. K. (2007). *Burkholderia phymatum* is a highly effective nitrogen-fixing symbiont of Mimosa spp. and fixes nitrogen ex planta. *New Phytologist*, 173: 168-180.

Galván, M. Z., Menéndez-Sevillano, M., De Ron, A. M. et al. (2006). Genetic diversity among wild common beans from Northwestern Argentina based on morphoagronomic and RAPD data. *Genetic Resources and Crop Evolution*, 53: 891-900.

García-Fraile, P., Mulas-García, D., Peix, A., Rivas, R., González-Andrés, F. and Velázquez, E. (2010). *Phaseolus vulgaris* is nodulated in northern Spain by *Rhizobium leguminosarum* strains harboring two *nodC* alleles present in American *Rhizobium etli* strains: biogeographical and evolutionary implications. *Canadian Journal of Microbiology*, 56: 657-666.

Gaunt, M. W., Turner, S. L., Rigottier-Gois, L., Lloyd-Macgilp, S. A. and Young, J. P. W. (2001). Phylogenies of *atpD* and *recA* support the small subunit rRNA-based classification of rhizobia. *International Journal of Systematic and Evolutionary Microbiology*, 51: 2037-2048.

Gepts, P., Osborn, T., Rashka, K. et al. (1986). Phaseolin-protein variability in wild forms and landraces of the common bean

(*Phaseolus vulgaris*): evidence for multiple centers of domestication. *Economic Botany*, 40: 451-468.

Gepts, P., Debouck, D. G. (1991). Origin, domestication and evolution of the common bean (*Phaseolus vulgaris* L.). In: van Schoonhoven, A. and Voysest, Om (eds.) Common Beans: Research for Crop Improvement, CAB Intrernational, Wallingford, UK and CIAT (Centro Internacional de Agricultura Tropical), Cali, pp. 7-53.

Gioia, T., Logozzo, G., Attene, G. et al. (2013). Evidence for introduction bottleneck and extensive inter-gene pool (Mesoamerica x Andes) hybridization in the European common bean (*Phaseolus vulgaris* L.) germplasm. *PLoS ONE*, 8:e75974.

González, A. M., Yuste-Lisbona, F., Lores, M. et al. (2014). Dissecting the genetic basis of popping ability in nuña bean, an ancient cultivar of common bean. *Euphytica*, 196: 349-363.

Gough, C. and Cullimore, J. (2011). Lipo-chitooligosaccharide signaling in endosymbiotic plant-microbe interactions. *Molecular Plant Microbe Interactions*, 24: 867-878.

Graham, P. H. (2008). Ecology of the root-nodule bacteria of legumes. In: Nitrogen-fixing Leguminous. Dilworth, M. J., James, E. K., Sprent, J. I. and Newton, W. E. (eds.). Springer, Dordrecht, The Netherlands, pp 23-43.

Hamblin, J. and Zimmermann, M. J O. (1986). Breeding common bean for yield in mixtures. *Plant Breeding Reviews*, 4: 245-270.

Hardarson G (1993). Methods for enhancing symbiotic nitrogen fixation. *Plant and Soil*, 152: 1-17.

Hammer, K. (1984). The domestication syndrome. *Kulturpflanze*, 32: 11-34.

Jarvis, D. I. and Hodgkin, T. (1999). Wild relatives and crop cultivars: detecting natural introgression and farmer selection of new genetic combinations in agro-ecosystem. *Molecular Ecology*, 8: S159-S173.

Jordan, D. C. (1984). Family III. Rhizobiaceae. In: Bergey's Manual of Systematic Bacteriology Vol. I. Krieg, N. R. and Holt, J. G. (eds.). Williams and Wilkins Co, Baltimore, USA, pp. 234-242.Kaplan, L. and Kaplan, L. N. (1988). *Phaseolus* in archaeology. In: Gepts PL (ed.)

Genetic Resources of *Phaseolus beans*. Kluwer Academic Publishers, Dordrecht, pp. 125-142.

Koenig, R. L. and Gepts, P. (1989). Allozyme diversity in *Phaseolus vulgaris* further evidence for two major centers of genetic diversity. *Theoretical and Applied Genetics*, 78: 809-817.

Koenig, R., Singh, S. P. and Gepts, P. (1990). Novel phaseolin types in wild and cultivated common bean (*Phaseolus vulgaris*, Fabaceae). *Economic Botany*, 44: 50-60.

Koinange, E. M. K., Singh, S. P. and Gepts, P. (1996). Genetic control of the domestication syndrome in common bean. *Crop Science*, 36: 1037-1045.

Kornegay, J., White, J. W. and Ortíz de la Cruz, O. (1992). Growth habit and gene pool effects on inheritance of yield in common bean. *Euphytica*, 62: 171-180,

Kornegay, J., White, J. W., Dominguez, J. R. et al. (1993). Inheritance of photoperiod response in Andean and Mesoamerican common beans. *Crop Science*, 33: 977-984.

Ladizinsky, G. (1985). Founder effect in crop-plant evolution. *Economic Botany*, 39: 191-199.

Latati, M., Bargazb, A., Belarbia, B. et al. (2016). The intercropping common bean with maize improves the rhizobial efficiency, resource use and grain yield under low phosphorus availability. *European Journao of Agronomy*, 72: 80-90.

López-López, A., Rogel-Hernández, M. A., Barois, I., Ortiz Ceballos, A. I., Martínez, J., Ormeño-Orrillo, E. and Martínez-Romero, E. (2012). *Rhizobium grahamii* sp. nov., from nodules of *Dalea leporina*, *Leucaena leucocephala* and *Clitoria ternatea*, and *Rhizobium mesoamericanum* sp. nov., from nodules of *Phaseolus vulgaris*, siratro, cowpea and *Mimosa pudica*. *International Journal of Systematic and Evolutionary Microbioloby*, 62: 2264-2271.

Mangelsdorf, P. C. and Smith, C. E. (1949). A discovery of remains of primitive maize in New Mexico. *Journal of Heredity*, 40: 39-43.

Martínez-Aguilar, L., Salazar-Salazar, C., Méndez, R. D., Caballero-Mellado, J., Hirsch, A. M., Vásquez-Murrieta, M. S. and Estrada de los

Santos, P. (2013). *Burkholderia caballeronis* sp. nov., a nitrogen fixing species isolated from tomato (*Lycopersicon esculentum*) with the ability to effectively nodulate *Phaseolus vulgaris*. Antonie Van Leeuwenhoek. 104: 1063-1071.

Martínez-Romero, E., Segovia, L., Mercante, F. M., Franco, A. A., Graham, P. and Pardo, M. A. (1991). *Rhizobium tropici*, a novel species nodulating *Phaseolus vulgaris* L. beans and *Leucaena* sp. trees. *International Journal of Systematic Bacteriology*, 41: 417-426.

Menéndez Sevillano, M. C. (2002). Estudio y conservación del germoplasma silvestre y primitivo de *Phaseolus vulgaris* L. en el Noroeste de Argentina. Tesis Doctoral. Universidad de Santiago de Compostela, España. [Study and conservation of the wild and primitive germplasm of *Phaseolus vulgaris* L. in the Northwest of Argentina. PhD Dissertation. University of Santiago de Compostela, Spain].

Michiels, J., Dombrecht, B., Vermeiren, N., Xi, C., Luyten, E. and Vanderleyden, J. (1998). *Phaseolus vulgaris* is a non-selective host for nodulation. *FEMS Microbiology and Ecology*, 26: 193-205.

Mnasri, B., Mrabet, M., Laguerre, G., Aouani, M. E. and Mhamdi, R. (2007). Salt-tolerant rhizobia isolated from a Tunisian oasis that are highly effective for symbiotic N_2-fixation with *Phaseolus vulgaris* constitute a novel biovar (bv. mediterranense) of *Sinorhizobium meliloti*. Arch. Microbiol. 187: 79–85.

Mnasri, B., Saïdi, S., Chihaoui, S. A., and Mhamdi, R. (2012). *Sinorhizobium americanum* symbiovar mediterranense is a predominant symbiont that nodulates and fixes nitrogen with common bean (*Phaseolus vulgaris* L.) in a Northern Tunisian field. *Systematic and Applied Microbiology*, 35: 263-269.

Mnasri, B., Liu, T. Y., Saidi, S., Chen, W.F., Chen, X. W., Zhang, X. X. and Mahmdi, R. (2014). *Rhizobium azibense* sp. nov., a nitrogen fixing bacterium isolated from root-nodules of *Phaseolus vulgaris*. *International Journal of Systematic and Evolutionary Microbiology*, 64: 1501-1506.

Moulin, L., Munive, A., Dreyfus, B. and Boivin-Masson, C. (2001). Nodulation of legumes by members of the β-subclass of Proteobacteria. *Nature*, 411: 948-950.

Mulas, D., García-Fraile, P., Carro, L., Ramírez-Bahena, M. H., Casquero, P., Velázquez, E. and González-Andrés, F. (2011). Distribution and efficiency of *Rhizo-bium leguminosarum* strains nodulating *Phaseolus vulgaris* in northern Spanish soils: Selection of native strains that replace conventional N fertilization. *Soil Biology and Biochemistry*, 43: 2283-2293.

Oldroyd, G. E. (2013). Speak, friend, and enter: signalling systems that promote beneficial symbiotic associations in plants. *Nature Review Microbiology* 11: 252-263.

Peix, A., Ramírez-Bahena, M. H., Velázquez, E. and Bedmar, E. J. (2015). Bacterial associations with legumes. *Critical Review in Plant Sciences*, 34: 17-42.

Pérez-Ramírez, N.O., Rogel, M. A., Wang, E., Castellanos, J. Z. and Martínez-Romero, E. (1998). Seeds of *Phaseolus vulgaris* bean carry *Rhizobium etli*. *FEMS Microbiology and Ecology*, 26: 289-296.

Ramírez-Bahena, M. H., García-Fraile, P., Peix, A., Valverde, A., Rivas, R., Igual, J. M., Mateos, P. F., Martínez-Molina, E. and Velázquez, E. (2008). Revision of the taxonomic status of the species *Rhizobium leguminosarum* (Frank 1879) Frank 1889 AL, *Rhizobium phaseoli* Dangeard 1926 AL and *Rhizobium trifolii* Dangeard 1926 AL. *R. trifolii* is a later synonym of *R. leguminosarum*. Reclassification of the strain *R. leguminosarum* DSM 30132 (=NCIMB 11478) as *Rhizobium pisi* sp. nov. *International Journal of Systematic and Evolutionary Microbiology*, 58: 2484-2490.

Ramos, L. M. G. and Boddey, R. M. (1987). Yield and nodulation of *Phaseolus vulgaris* and the competitiveness of an introduced *Rhizobium* strain: effects of lime, mulch and repeated cropping. *Soil Biology and Biochemistry*, 19: 171-177.

Ribeiro, R. A., Rogel, M. A., López-López, A., Ormeño-Orrillo, E., Barcellos, F. G., Martínez, J., Thompson, F. L., Martínez-Romero, E., and Hungria, M. (2012). Reclassification of *Rhizobium tropici* type A strains as *Rhizobium leucaenae* sp. nov. *International Journal of Systematic and Evolutionary Microbiology*, 62: 1179-1184.

Rivas, R., García-Fraile, P. and Velázquez, E. (2009). Taxonomy of bacteria nodulating legumes. *Microbiology Insights*, 2: 251-269.

Román-Ponce, B., Zhang, Y. J., Vásquez-Murrieta, M. S., Sui, X. H., Chen, W. F., Padilla, J. C. A., Guo, X. W., Gao, J. L, Yan, J., Wei-Ge, H. and Wang, E. T. (2016). *Rhizobium acidisoli* sp. nov., isolated from root nodules of *Phaseolus vulgaris* in acid soils in Mexico. *International Journal of Systematic and Evolutionary Microbiology*, 66: 398-406.

Santalla, M., De Ron, A. M. and Escribano, M. R. (1994). Effect of intercropping bush bean populations with maize on agronomic traits and their implication for selection. *Field Crops Research*, 36: 185-189.

Santalla, M., De Ron, A. M. and Casquero, P. A. (1995). Nutritional and culinary quality of bush bean populations intercropped with maize. *Euphytica*, 84: 57-65.

Santalla, M., Fueyo, M.A., Rodiñom A. P. et al. (1999). Breeding for culinary and nutritional quality of common bean (*Phaseolus vulgaris* L.) in intercropping systems with maize (*Zea mays* L.). *Biotechnoly Agronomy Society and Environment*, 3: 225-229.

Santalla, M., Amurrio, J. M., Rodiño, A. P. and De Ron, A. M. (2001). Variation in traits affecting nodulation of common bean under intercropping with maize and sole cropping. *Euphytica*, 122: 243-255.

Santalla, M., Rodiño, A. P. and De Ron, A. M. (2002). Allozyme evidence supporting southwestern Europe as a secondary center of genetic diversity for common bean. *Theoretical and Applied Genetics*, 104: 934-944.

Santalla, M., Menéndez-Sevillano, M. C., Monteagudo, A. B. et al. (2004). Genetic diversity of Argentinean common bean and its evolution during domestication. *Euphytica*, 135: 75-87.

Schmutz, J., McClean, P. E., Mamidi, S. et al. (2014). A reference genome for common bean and genome-wide analysis of dual domestications. *Nature Genetics*, 46: 707-713.

Segovia, L., Young, J. P. W. and Martínez-Romero, E. (1993). Reclassification of American *Rhizobium leguminosarum* biovar phaseoli type I strains as *Rhizobium etli* sp. nov. *International Journal of Systematic and Evolutionary Microbiology*, 43: 374-377.

Sessitsch, A., Hardarson, G., Akkermans, A. D. and de Vos, W. M. (1997). Characterization of *Rhizobium etli* and other *Rhizobium spp.* that nodulate *Phaseolus vulgaris* L. in an Austrian soil. *Molecular Ecology*, 6: 601-608.

Shamselding, A., Abdelkhalek, A. and Sadowsky, M. J. (2017). Recent changes to the classification of symbiotic, nitrogen-fixing, legume-associating bacteria: a review. *Symbiosis* 71: 91-109.

Silva, C., Vinuesa, P., Eguiarte, L. E., Martínez-Romero, E. and Souza, V. (2003). *Rhizobium etli* and *Rhizobium gallicum* nodulate common bean (*Phaseolus vulgaris*) in a traditionally managed milpa plot in Mexico: population genetics and biogeographic implications. *Applied and Environmental Microbiology*, 69: 884-893.

Singh, S.P., Gepts, P. and Debouck, D. G. (1991). Races of common bean (*Phaseolus vulgaris* Fabaceae). *Economic Botany*, 45: 379-396.

Singh, S. P., Molina, A. and Gepts, P., (1995). Potential of wild common bean for seed yield improvement of cultivars in the tropics. *Canadian Journal of Plant Sciences*, 75: 807-813.

Singh, S. P. (2001). Broadening the genetic base of common bean cultivars: a review. *Crop Science*, 41: 1659-1675.

Smartt, J. (1988). Morphological, physiological and biochemical changes in *Phaseolus* beans under domestication. In: Gepts, P. (ed.). Genetics Resources of *Phaseolus* Beans: Their Maintenance, Domestication, Evolution and Uilization. Kluwer, Dordrecht, pp 543-560.

Sprent, J. I. (2007). Evolving ideas of legume evolution and diversity: A taxonomic perspective on the occurrence of nodulation. *New Phytologist*, 174: 11-25.

Sprent, J. I., Ardley, J. and James, E. K. (2017). Biogeography of nodulated legumes and their nitrogen-fixing symbionts. *New Phytologist*, 215: 40-56.

Stem, W. R. (1993). Nitrogen fixation and transfer in intercropping. *Field Crops Research*, 34: 335-356.

Talbi, C., Delgado, M. J., Girard, L., Ramírez-Trujillo, A., Caballero-Mellado, J. and Bedmar, E. J. (2010). *Burkholderia phymatum* strains capable of nodulating *Phaseolus vulgaris* are present in Moroccan soils. *Applien and Environmental Microbiology*, 76: 4587-4591.

Tohme, J., Toro, O., Vargas, J. and Debouck, D. G. (1995). Variability in Andean ñuña common beans (*Phaseolus vulgaris*, Fabaceae). *Economic Botany*, 49:78-95.

Tohme, J., González, D. O., Beebe, S. and Duque, M. C. (1996). AFLP analysis of gene pools of a wild bean core collection. *Crop Science*, 36: 1375-1384.

Valverde, A., Igual, J. M., Peix, A., Cervantes, E. and Velázquez, E. (2006). *Rhizobium lusitanum* sp. nov. a bacterium that nodulates *Phaseolus vulgaris*. *International Journal of Systematic and Evolutionary Microbiology*, 56: 2631-2637.

Vandamme, P., Goris, J., Chen, W. M., de Vos, P. and Willems, A. (2002). *Burkholderia tuberum* sp. nov. and *Burkholderia phymatum* sp. nov., nodulate the roots of tropical legumes. *Systematic and Applied Microbiology*, 25: 507-512.

van Schoonhoven, A. and Cardona, C. (1982). Low levels of resistance to the Mexican bean weevil in dry beans. *Journal of Economic Entomology*, 75: 567-569.

van Schoonhoven, A., Cardona, C. and Valor, J. (1983). Resistance to the bean weevil and the Mexican bean weevil (Coleoptera: Bruchidae) in non-cultivated common bean accessions. *Journal of Economic Entomology*, 76: 1255-1259.

Velázquez, E., García-Fraile, P., Ramírez-Bahena, M. H., Peix, A. and Rivas R. (2010). Proteobacteria forming nitrogen fixing symbiosis with higher plants. In: Sezenna M. L. (ed.) Proteobacteria: Phylogeny, metabolic diversity and ecological effects. Nova Science Publishers Inc. New York, USA, pp. 37-56.

Vigouroux, Y., MacMullen, M., Hittinger, C. T. et al. (2002). Identifying genes of agronomic importance in maize by screening microsatellites for evidence of selection during domestication. *Proceedings of the National Academy of Sciences* USA 99: 9650-9655.

Vlasova, A., Capella-Gutiérrez, S., Rendón-Anaya, M. et al. (2016). Genome and transcriptome analysis of the Mesoamerican common bean and the role of gene duplications in establishing tissue and temporal specialization of genes. *Genome Biology*, 17:32. 10.1186/s13059-016-0883-6.

Wang, F., Wang, E. T., Wu, L. J., Sui, X. H., Li, Y. Jr. and Chen, W. X. (2011). *Rhizobium vallis* sp. nov., isolated from nodules of three leguminous species. *International Journal of Systematic and Evolutionary Microbiology*, 61: 2582-2588.

White, J.W., Kornegay, J., Castillo, J. et al. (1992). Effect of growth habit on yield of large-seeded bush cultivars of common bean. *Field Crops Research*, 29: 151-161.

In: Legumes for Global Food Security ISBN: 978-1-53612-265-7
Editors: Alfonso Clemente et al.

Chapter 4

TRICHODERMA SPECIES AS BIOCONTROL AGENTS IN LEGUMES

Sara Mayo[1], Santiago Gutiérrez[1], Rosa E. Cardoza[1], Rosa Hermosa[2], Enrique Monte and Pedro A. Casquero[1,*]

[1]Research Group of Engineering and Sustainable Agriculture, Universidad de León, León, Spain

[2]Spanish-Portuguese Centre for Agricultural Research, University of Salamanca, Salamanca, Spain

ABSTRACT

Trichoderma spp. (Teleomorph: *Hypocrea*) is a fungal genus that is found in the soil, and it is a secondary fast growing opportunistic invasive producer of chitinases, glucanases and proteases, and metabolites with antimicrobial activity. Many *Trichoderma* species are also well known as biocontrol agents of important phytopathogenic fungi. The primary mechanisms of biocontrol used by *Trichoderma* in direct confrontation

---

* Corresponding Author address. Dr Pedro A. Casquero, Research Group of Engineering and Sustainable Agriculture, Universidad de León, Av. Portugal 41, 24071 León, Spain. Email: pedro-casquero@unileon.es.

with pathogenic fungi are the mycoparasitism, antibiosis, and competition for nutrients with the pathogen. *Trichoderma* species have developed opportunistic mechanisms for their adaptation to abiotic stresses as well as for nutrient uptake and solute transport. *Trichoderma* is able to promote the growth of legumes. *Trichoderma* also produces some organic acids such as citric or fumaric acid that reduces soil pH and allows the solubilization of phosphates and other micronutrients such as iron, manganese and magnesium in legume crops. In the legumes, *Trichoderma* spp. produces several changes in the defence response of the plant, as the accumulation of phytoalexins and down- or up-regulation of defence-related genes expression.

Keywords: Biocontrol, Defence Response, Induction of Resistance, Mycoparasitism, Nutrients, Rhizosphere, *Trichoderma*

1. INTRODUCTION

Pulses belong to the Fabaceae or Leguminosae family; these plants are the world's third largest group of plant life. They are thought to have originated some 90 million years ago, with a diversification process beginning in the early Tertiary era. The Fabaceae family contains over 20,000 species and 700 genera, of which only some are categorised as leguminous plants, such as the genus *Phaseolus*, *Vicia*, *Cicer*, *Lens,* etc (Food and Agriculture Organization, 2016).

A plant is diseased when it is not functioning normally. The abnormal functioning of a plant generally leads to reduction in quantity and quality of yield. Disease is the result of an interaction among the plant and its environmental and it is often affected by biotic and abiotic factors by (microorganisms, humidity, temperature…) that are detected as signals for the activation of plant response mechanisms (American Phytopathological Society, 2005). These crops are affected by a wide diversity of fungal pathogens for example *Sclerotinia* spp., *Fusarium* spp., *Pythium* spp., *Botrytis* spp, *Rhizoctonia* spp., causing important economic losses in these crops.

A form of control to disease is the application of chemical synthesis fungicides. Its application on the seed or directly to the soil can be effective against fungi that affect the crops during or shortly after germination (Beebe, Corrales, Schoonhoven, & Voysest, 1991) because they reduce its incidence and improve the emergence of plants (Valenciano, Casquero, & Boto, 2004). However, applications with fungicides aimed at avoiding damage caused by fungi that cause root rot or yellowing and wilting are often ineffective and usually impracticable due to the large volume of soil to which they should be directed.

Actually the number of authorized products by protection plant has been reduced in order to ensure food safety and its sustainable in the long term. It is therefore proposed to prioritize non-chemical methods in integrated production, organic farming, and others.

As a strategy to control legume infectious diseases, mainly those caused by fungi, the use of biocontrol agents can reduce the negative effects of plant pathogens and they also can promote positive responses in the plant (Shoresh, Harman, & Mastouri, 2010). Biocontrol agents are perceived to have specific advantages over synthetic fungicides, including fewer non-target and environmental effects, efficacy against fungicide-resistant pathogens, reduced probability of resistance development and use in organic farming situations where synthetic fungicides are restricted (Brimner & Boland, 2003).

The genera including strains belonging to bacterial genera such as *Agrobacterium*, *Pseudomonas*, *Streptomyces* and *Bacillus*, and fungal genera such as *Gliocladium*, *Trichoderma*, *Ampelomyces*, *Candida* and *Coniothyrium*., are beneficial organisms that have shown good efficiency as biocontrol agents against pathogenic microorganisms (Vinale et al., 2008).

2. *TRICHODERMA* SPP.

Trichoderma spp. (Teleomorph: *Hypocrea*) is a fungal genus that is found in the soil, and it is a secondary fast growing opportunistic invasive

(Mayo, Cominelli, et al., 2016) producer of chitinases, glucanases and proteases, and metabolites with antimicrobial activity (Lorito, Woo, Harman, & Monte, 2010). Many *Trichoderma* species are also well known as biocontrol agents of important phytopathogenic fungi. The primary mechanisms of biocontrol used by *Trichoderma* in direct confrontation with pathogenic fungi are the mycoparasitism, antibiosis, and competition for nutrients with the pathogen (Harman, Howell, Viterbo, Chet, & Lorito, 2004). *Trichoderma* species colonize the root surface and cause substantial changes in plant metabolism (Shoresh et al., 2010). The physical interaction between *Trichoderma* and plants is limited to the first cell layer of the epidermis and the root bark. In addition, *Trichoderma* biocontrol strains are able to induce the expression of genes involved in defence response and also to promote plant growth, root development and nutrient uptake (Hermosa, Viterbo, Chet, & Monte, 2012) (Figure 1).

Trichoderma spp. is recognized for their important benefits to agriculture such as its ability to protect crops against diseases (Benítez, Rincón, Limón, & Codón, 2004) and increase crop yield under field conditions (Harman et al., 2004). Most species of *Trichoderma* have been linked to biocontrol and biotechnological applications (Monte, 2001), and the versatility of *Trichoderma* strains to suppress diseases caused by pathogens (Howell, 2003). Since *Trichoderma* strains grow and proliferate best when there are abundant healthy roots, they have evolved numerous mechanisms of action both to attack other fungi and to enhance plant and root growth (Benítez et al., 2004).

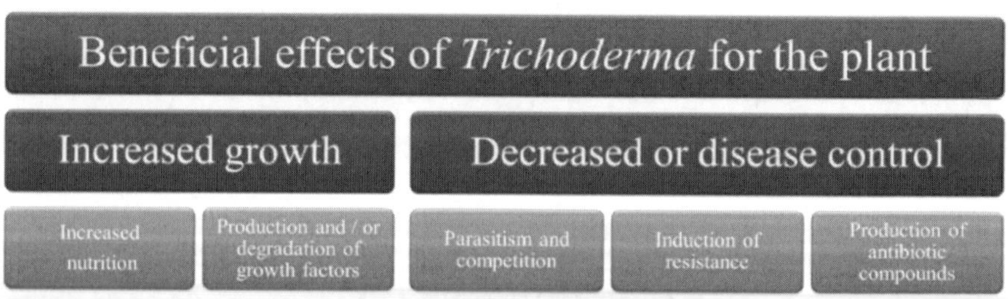

Figure 1. Beneficial effect of *Trichoderma* for the plant.

2.1. Promotion of Plant Growth

As rhizosphere colonizers, *Trichoderma* species have developed opportunistic mechanisms for their adaptation to abiotic stresses as well as for nutrient uptake and solute transport. In the plant, these processes are facilitated by the induction of cell wall extension and expansion, secondary root development, lateral root hair production and a higher photosynthetic rate (Shoresh et al., 2010; Hermosa et al., 2013).

In the culture of peanut, *Arachis hypogaea*, it has been checked the effect of *Trichoderma* inoculum concentration on plant emergence with and without a fungicide that could be used together with the antagonist species to improve seed protection and emergence. Considering that the fungicide alone is effective only for a short time and it does not spread through the soil with the root system, the use of both chemical and biological agents could increase the protection of seeds and roots (Rojo, Reynoso, Ferez, Chulze, & Torres, 2007).

In other work, when the peanuts were inoculated with four *Trichoderma* (*T. harzianum, T. viride, T. virens, T. atroviride*), length of shoot was increased between 51.78% by *T. harzianum* and 10.48% by *T. atroviride* respect to control plant; weight of the shoot was increased among 80.38% (*T. harzianum*) and 14.04% (*T. atroviride*); the weight of the roots was enhanced 64.42% and the number of nodules per plant was 30.97% respect to control plant when they inoculated with *T. harzianum. Trichoderma* spp. can be a useful tool that improves the microbial community in the rhizosphere, which result in an enhancement of plant growth and development. In this study *Trichoderma* stimulated early growth in peanut plants, increasing the yield (Kamaruzzaman, Rahman, Islam, Ahmad, & Ahmad, 2016).

Bean plants, *P. vulgaris*, in contact with some *Trichoderma*, showed an increased size when the pathogen was not present. Other authors also observed that *T. harzianum* was able to promote the growth of common bean plants in comparison to plants grown in its absence (Pereira et al., 2014). However, plants treated with *T. harzianum* did not decrease their size in the presence of *R. solani* (Mayo et al., 2015). In other work, when bean plants were in contact with *T. velutinum* isolate, they increased respect to control plants 4.75% their diameter of hypocotyl, 10.75% their root system length, 4.27% and 5.51% in dry weight of aerial parts and root system, respectively. When plants were infected with the pathogen *R. solani*, the action of *T. velutinum* T028 caused an increased of the diameter of the hypocotyl in 8.76%, 21.15% in the length of root system, and 11.05% and 3.43% in dry weight of aerial parts and root system, respectively, respect to the control plant with the pathogen (Mayo et al., 2016). These data are in agreement with those from Yedidia and coworkers reporting a much stronger effect on cucumber plants treated with *T. harzianum*, which increased by a 75.0% the root length, 95.0% aerial parts, 80.0% dry weight and 80.0% the size of the blade relative to the untreated control (Yedidia, Srivastva, Kapulnik, & Chet, 2001). *T. harzianum* was able to promote the growth of common bean plants in comparison to plants grown in its absence (Pereira et al., 2014). In other study, it was evaluated the effect of farnesol on the development of bean plants. This compound that in fungi is presumably produced from farnesyl pyrophosphate, is a signalling molecule for self-regulation that by accumulating in the extracellular space generates a response across local fungal population. The plants were grown under controlled conditions in hydroponic culture for 14 days with different concentrations of farnesol (10 μM, 100 μM, 1000 μM, 2000 μM and 5000 μM) in the nutrition solution and they were evaluated by assessing the wet weight and dry weights of the aerial parts and root systems. The results showed a negative effect on growth at concentrations of 10 μM and 100 μM farnesol, which could be related to abscisic acid synthesis. However, with 2 mM of farnesol, bean plants showed increased development of aerial parts and root systems (Mayo, Izquierdo, et al., 2016).

In other study, Dubey and co-authors (2007) reported *Trichoderma* spp. signicantly reduced the wilt incidence in chickpea plants, *Cicer arietinum*, that were inoculated with *Fusarium oxysporum* f. sp. *Ciceris*. The incidence was reduced a 65.85% respect to control plants. Also, the presence of *Trichoderma* increased the germination of this culture respect to control plants with values between 13.36% and 8.95%, the weight of the shoot was increased 31.31% and the weight of the root was a 41.30% higher that in control plants.

Trichoderma produces some organic acids such as citric or fumaric acids that reduces soil pH and allows the solubilization of phosphates and other micronutrients such as iron, manganese and magnesium (Benítez et al., 2004; Harman et al., 2004). It has been observed that *T. harzianum* reduces the amount of iron, manganese, zinc and copper in the leaf biomass of beans, but increases the content of copper in its seeds (Öğüt & Er, 2006). On the other hand, there are some *in vitro* studies which indicate that *T. harzianum* and other isolates of *Trichoderma* could solubilize iron (III) oxide, manganese (IV) oxide, zinc and phosphates, which are highly insoluble compounds or with low solubility, owing to chelation processes and oxidation-reduction activity (Altomare, Norvell, Björkman, & Harman, 1999). The increment of all those nutrients, in particular phosphorus, could favor the growth of the plants, which are in contact with soils where this fungal species is present. It has been shown that *T. atroviride* produces and degrades indoleacetic acid, which when is produced in combination with ethylene by the microorganisms present in the rhizosphere causes an increase in plant growth (Gravel, Antoun, & Tweddell, 2007).

Chagas and co-authors (2016) inoculated cowpea, *Vigna unguiculata*, with some *Trichoderma* isolates. An increase in dry matter and phosphorous amounts in the aerial parts was observed in comparison to the control treatment. This effect might be explained by the phosphate solubilization ability of the biological control agents. Various factors might influence the *Trichoderma* phosphate solubilization ability, including the availability of carbon and nitrogen, the type of cultivated plants (Grayston, Vaughan, & Jones, 1997), and the type of phosphate that is solubilized (Barroso & Nahas, 2005). Similar results were also reported by Kapri and Tewari (2010), who inoculated *Trichoderma* into chickpea cultures and reported an increase in the phosphate solubilization ability *in vitro*, with a significant effect on biomass production. These works highlighted the potential of *Trichoderma* spp. as plant growth promoters in association with their phosphate solubilizing and indole acetic acid synthesis abilities.

According to other study the increase in biomass production related to growth hormones or analogs production is another mechanism by which strains of *Trichoderma* spp. can improve the growth of plants (Hoyos-Carvajal, Orduz, & Bissett, 2009).

2.2. Mycoparasitism

Mycoparasitism consists in the recognition of the fungus, attacking it and penetrating it with the purpose to cause its death. This process involves some different phases. First of all, *Trichoderma* locates the pathogen without previous contact, beginning to enlarge towards the pathogen by tropism (Chet, Harman, & Baker, 1981; Lu et al., 2004). During this process, *Trichoderma* secretes some enzymes that hydrolyze the cell wall of the pathogen (Howell, 2003; Woo, Scala, Ruocco, & Lorito, 2006). It has been studied that *Trichoderma* releases an extracellular exochitinase (Brunner et al., 2003) that might cause the liberation of some oligomers from the fungus, which could induce the expression of endochitinases that would diffuse and would start to attack to the pathogen even before the physical contact had happened. Some enzymes belonging to these fungi have been purified and used for biocontrol. When they have been assessed, they have showed antifungal activity and have controlled a large number of pathogens, such as *Fusarium, Rhizoctonia, Alternaria, Ustilago, Venturia* and *Colletotrichum* (Lorito et al. 1993; Lorito, Hayes, Di Pietro, Woo, & Harman, 1994).

A major part of the *Trichoderma* antifungal system consists of a number of genes encoding an astonishing variety of secreted lytic enzymes (Sanz et al., 2004) including endochitinases, N-acetyl-β-glucosaminidases, chitin 1,4-β-chitobiosidases, proteases, glucan β-1,3-glucosidases, glucan β-1,6-glucosidases, glucan α-1,3-glucosidases, lipases, xylanases, mannanases, pectinases, pectin lyases, amylases, phospholipases, RNAses, DNAses, etc. Some of these proteins have been purified and their corresponding genes have been cloned and characterized by our group: protease PRA1 (Suarez, Rey, Castillo, Monte, & Llobell, 2004), chitinases

CHIT36 and CHIT37 (Viterbo et al., 2002), α-glucanases AGN13.1 (Ait-Lahsen et al., 2001) and AGN13.2 (Sanz, Montero, Redondo, Llobell, & Monte, 2005), and β-1,6-glucanases BGN16.2 and BGN16.3 (Montero, Sanz, Rey, Llobell, & Monte, 2007; Montero, Sanz, Rey, Monte, & Llobell, 2005).

The direct confrontation assays were used to verify the ability of *Trichoderma* spp. to overgrow the pathogen and its f mycoparasitim potential. In a study, the percentage of overgrow oscillated between 72.77 and 14.63%, according to the species of *Trichoderma* (Mayo et al., 2015).

Some *Trichoderma* spp. are selected because of their mycoparasitic mechanism but the most efficient biocontrol strains display, simultaneous or sequentially, more than one biocontrol strategy (Howell, 2003). *Trichoderma* spp. can also exert marked antimicrobial activity (Vizcaino et al., 2005) due to the production of blends of secondary metabolites (Cardoza et al., 2005; Reino, Guerrero, Hernández-Galán, & Collado, 2007) of which we have studied the pyrones [6-pentyl-2H-pyran-2-one (6-PP) derivatives] (Keszler et al., 2000) and peptaibols (Vizcaino et al., 2005). Currently, better knowledge about *Trichoderma* has facilitated its use in biocontrol as whole microorganisms, able to be monitorized in natural environments (Hermosa, Grondona, Díaz-Mínguez, Iturriaga, & Monte, 2001; Rubio, Hermosa, Keck, & Monte, 2005), as enzyme formulations (Benítez et al., 2004) or as sources of genes for transgenic plant development (M Lorito et al., 1998). Since the early description of the capacity of *Trichoderma* to increase plant biomass production (Chang, Chang, Baker, Kleifeld, & Chet, 1986), several new general mechanisms for both biocontrol and plant growth increase have been demonstrated and it is now clear that there must be hundreds of separate genes and gene products involved in these processes.

2.3. Competition for Nutrients

Competition between *Trichoderma* and pathogens would be established with the purpose to get more nutrients, oxygen, light, (Paulitz,

1990). *Trichoderma* is an excellent competitor for space and nutritional resources. It appears in almost all soils and in habitats that contains high amounts of organic matter. In those places, it would be an excellent decomposer of plant and fungal material. Moreover, some species of the genus *Trichoderma* show great metabolomic versatility that allows them to grow using a wide range of nitrogen and carbon sources. Furthermore, *Trichoderma* has the ability to colonize the rhizosphere, and this skill might be essential for being used as an excellent biological control agent (Howell, 2003).

Genes related to the capacity of *Trichoderma* to compete in soil and rhizosphere colonization (Chacón et al., 2007) were also described: the first oligopeptide transporter analyzed functionally in filamentous fungi (Vizcaíno et al., 2006) and two heat shock proteins that have been overexpressed in a biocontrol strain of *Trichoderma* and in plant, improving their resistance to abiotic stresses (Montero, Sanz, Rey, Monte, & Llobell, 2005; Montero, Sanz, Rey, Llobell, & Monte, 2007).

2.4. Production of Antibiosis Compounds

Trichoderma produces a large number of secondary metabolites with biological activity (Ghisalberti & Sivasithamparam, 1991; Sivasithamparam & Ghisalberti, 1998) that includes some natural chemical compounds related with its survival, its competition ability against organisms, as well as its symbiosis, its transport, its differentiation ... (Demain & Fang, 2000). Furthermore, antibiotics, which could inhibit microbial growth, are also included in this group.

In the antifungal assay on membranes, carried out by Mayo and co-authors (2015), where it was determined the ability of *Trichoderma* isolates to produce metabolites with inhibitory activity against *R. solani* the percentages of inhibition ranged between 86% and 58%. In others assays the percentage of inhibition in membrane assays using *T. virens* T59 and *T. atroviride* T11 range between 100% and 84.7%, respectively (Campelo et al., 2010).

2.5. Induction of Resistance

The relationships established between plant and microorganisms are very diverse, when a plant is exposed to a pathogenic microorganism, the production of molecules associated to salicylic acid is increased, being this a systemic acquired resistance (SAR) response. The response of plants against non-pathogenic microorganisms is different, resulting in activation of signalling cascades that are dependent on jasmonic acid and ethylene, such as hydroperoxide lyase, peroxidase and phenylalanine ammonia lyase, all of which belong to an induced systemic resistance (ISR) response (Druzhinina et al., 2011). Other responses result in a rapid cell death in infected tissues, then plants activate the hypersensitive response that involves the accumulation of salicylic acid, reactive oxygen species and an increased the influx of Ca^{2+} (Guerrero-González et al., 2011).

In the tripartite interaction of legume plants with a pathogen and a biocontrol *Trichoderma* species, several changes are produced in the plant, such as the increase in phenolic acid and lignin, accumulation of phytoalexins and down- or up-regulation of defence-related genes expression (Guerrero-González et al., 2011; Mayo et al., 2015) Different categories of defence-related genes whose expression is modulated by biotic stresses have been described in bean plants interacting with pathogen and non-pathogenic microorganisms (Mayo et al., 2015). Several works have studied the expression of defence-related genes in the plants when they have in contact with *Trichoderma* and/or pathogen.

WRKY transcription factors have been involved in the regulation of plant defence gene expression (Rushton & Somssich, 1998; Singh, Foley, & Oñate-Sánchez, 2002). Thus, *WRKY33* has a role in biotic stress defence, where it regulates the balance between necrotrophic and biotrophic pathogen responses (Birkenbihl, Diezel, & Somssich, 2012; Lippok et al., 2007; Pandey & Somssich, 2009). Previous studies have pointed out the involvement of *Arabidopsis* WRKY transcription factors in regulating the expression of *PR* genes by direct binding (W. Chen et al., 2002; Kim, Fan, & Chen, 2006). A rapid pathogen-induced *WKRY33* expression did not require salicylic acid signalling but a down-regulation of this gene involved a direct activation of jasmonic acid (Bakshi & Oelmüller, 2014). In the work by Mayo and co-workers (2016), when bean plants were in contact *T. velutinum* T028 without pathogen, *WRKY33* expression was significantly up-regulated while the *PR* genes expression (*PR2*, *PR3* and *PR4*) was significantly down-regulated compared to expression levels in plants without *Trichoderma* treatment. However, in the same work, when the pathogen *Rhizoctonia solani* was added to the substrate, expression of *WRKY33* was significantly down-regulated in plants with *Trichoderma* inoculation, while *PR2*, *PR3* and *PR4* were down-regulated.

WRKY33 is also involved in the regulation of the expression of genes modulated by components of the ethylene-signalling pathway. Expression of the *ERF1* and *ERF5* reached similar significant values either with or without *Trichoderma* and or *R. solani* in the substrate. *WRKY33* would act as a represor of *ERF1* and *ERF5* expression. Thus, when the expression of *WRKY33* is increased, expression of *ERF1* and *ERF5* is down-regulated (Mayo et al., 2016).

Other gene is *CH5b* encodes an endochitinase precursor and it is also related with the ethylene- signalling pathway. In previous works, it has been shown that, when this gene was over-expressed the *R. solani* symptoms were reduced in crops like *Nicotiana tabacum* and *Brassica napus* (Broglie et al., 1991). However, in the study by Mayo et al. (2016), when *P. vulgaris* plants were in contact with *R. solani,* the expression of this gene was down-regulated but not significantly, while treatment of these infected plants with *T. velutinum* resulted in its significant up-regulation. These results are in agreement with previous data, showing that the pathogen represses its expression, and the presence of *Trichoderma* induced it (Mayo et al., 2015).

PAL plays an important role in plant defence; it is involved in the biosynthesis of salicylic acid, which is related to plant systemic resistance (Chaman, Copaja, & Argandoña, 2003; Mauch-Mani & Slusarenko, 1996; Nugroho, Verberne, & Verpoorte, 2002). *PAL* gene expression is also regulated in response to pathogen infection. The presence of *T. velutinum* and *R. solani* in the soil resulted in a significant down-regulation of this gene compared with control plants (Mayo et al., 2016).

The *CNGC* genes can be related to early plant defence responses due to changes in ion flux, including H^+ and Ca^{2+} influx and K^+ and Cl^- efflux (Atkinson, Midland, Sims, & Keen, 1996). The up-regulation of *CNGC2* can confirm the importance of ion channels for the plant resistance response (Borges, Melotto, Tsai, & Caldas, 2012). *CNGC2* was down-regulated in plants treated with *T. velutinum* (Mayo et al., 2016).

HPL (hydroperoxide lyase) is involved in the production of antimicrobial and defence signalling oxylipins (Huang, Studart Witkowski, & Schwab, 2010; Noordermeer, Veldink, & Vliegenthart, 2001). In this study, the presence of *T. velutinum* and *R. solani*, resulted in a down-regulation of this gene expression respect to control plants. In the present case, after 45 days in contact with the fungus *T. velutinum* and/or *R. solani*, its expression was down-regulated, indicating that the plant identifies *Trichoderma* and *Rhizoctonia* as two invader organisms, and some of the mechanisms activated against the presence of both are similar, independently of the final response specifically activated in the plant by each one (Mayo et al., 2016).

hGS encodes a homoglutathione synthetase that is involved in response to oxidative stress. There is not much information about the behaviour of this gene in the plant. In the study of Mayo and coworkers (2016), when bean plants were in contact with *T. velutinum* and/or *R. solani*, expression of this gene was significantly up-regulated compared to control plants. In other studies, treatment of *Medicago truncatula* plants with compounds that release nitric oxide, a key signalling molecule in plants, induced expression of *GST* but not *hGS* in roots (Innocenti et al., 2007). Similarly, common bean plants treated with H_2O_2 showed up-regulation of *hGS* in nodules, whereas treatments with cadmium, sodium chloride, or jasmonic acid had no effect (Loscos, Matamoros, & Becana, 2008).

2.6. Production of Antibiotic Compounds: Changes in Plant Metabolism as Defence Response

When a plant is induced by exposure to a microorganism, starts the production of the biochemical products or physiological changes and activation of various metabolic pathways depends on the type and origin of these signalling natural products. Different secondary metabolites are synthesized after perception and recognition of the signals originating from plant or pathogenic microorganism elicitors created during the first steps of plant defence reactions (Boller & Felix, 2009; Grotewold, 2005; Veitch, 2009). Plants respond after the invasion of a phytopathogen or a biocontrol agent that activates disease resistance response to the invasion, by inducing the expression of defence genes (Mayo et al., 2016) and by the production of some secondary metabolites with antibacterial activity, such as phytoalexins (phenols, isoflavones, terpenes), and some substances that can block pathogen invasion and spread, such as lignin and callose (X. Chen, Qi, & Duan, 2015). Some plants do not produce phytoalexins when are in contact with pathogens, but release toxins that are normally stored as less toxic glycosides (Grayer & Kokubun, 2001).

Trichoderma species also produce plant hormones and solubilize minerals in the soil, which help to promote plant growth and suppress the disease (Kim et al., 2006).

During the *Trichoderma*-plant interaction various classes of metabolites could induce resistance such as proteins with enzymatic activity, low molecular weight compounds, related to the fungal or the plant cell wall, originated by the enzymatic activity of *Trichoderma* (Woo et al., 2006; Woo & Lorito, 2007) and other secondary metabolites that trigger plant defence mechanisms against the pathogen (Hermosa et al., 2012; Malmierca et al., 2014), by inducing the expression of pathogenesis–related (PR) proteins that reduce the diseases symptoms.

During the plant-*Trichoderma* interactions, the fungus participates actively in protecting and improving its ecological niche. Leucine-rich repeat (LRR)-containing proteins are signal receptors regulating plant development and defence (Afzal, Wood, & Lightfoot, 2008). In the work of Marra and co-authors (2006) observed that LRR proteins increased in bean leaves (*Phaseolus vulgaris*)interacting with *T. atroviride*, and that hydrophobins and ABC transporters were accumulated in the proteome of the fungus. Hydrophobins (Rosado et al., 2007) and ABC transporters (Ruocco et al., 2007) support the biocontrol activity of *Trichoderma* and its ability to colonize the roots. In a similar way, a *Trichoderma*-secreted swollen in (an expansin-like 5 protein) remarkably increased fungus plant root colonization efficiency and, due to a cellulose-binding domain was able to trigger defence responses in the plant and afforded pathogen protection, indicating that this domain might therefore be recognized by the plant as a microbe-associated molecular pattern (MAMP) in the *Trichoderma*-plant interaction (Brotman, Briff, Viterbo, & Chet, 2008). At least four classes of substances that elicit plant defence responses have been identified in *Trichoderma*: polysaccharide oligomers, enzymes, low molecular weight proteins and peptaibols. Some cell wall oligomers may act as elicitor molecules released by plants following pathogen attack (Woo et al., 2006), the overexpression of *Trichoderma* chitinase genes in tobacco plants generates innate defence responses and enhanced stress tolerance (Dana, Pintor-Toro, & Cubero, 2006), hydrophobin-like cysteine-rich low molecular weight secreted proteins Sm1 from *T. virens* and Epl1 from *T. atroviride* (Djonović, Pozo, Dangott, Howell, & Kenerley, 2006; Seidl, Marchetti, Schandl, Allmaier, & Kubicek, 2006) can also trigger ISR but, with the exception of peptaibols as elicitors of plant defence responses (Viterbo, Wiest, Brotman, Chet, & Kenerley, 2007), the role of secondary metabolites in this task remains unexplored.

References

Afzal, A. J., Wood, A. J. and Lightfoot, D. A. (2008). Plant receptor-like serine threonine kinases: Roles in signaling and plant defense. *Molecular Plant-Microbe Interactions*, *21(5),* 507–517. http://doi.org/10.1094/MPMI-21-5-0507.

Ait-Lahsen, H., Soler, A., Rey, M., de La Cruz, J., Monte, E. and Llobell, A. (2001). An antifungal exo-alpha-1,3-glucanase (AGN13.1) from the biocontrol fungus *Trichoderma harzianum*. *Applied and Environmental Microbiology, 67(12)*, 5833–5839. http://doi.org/10.1128/AEM.67.12.5833-5839.2001.

Altomare, C., Norvell, W. A., Björkman, T. and Harman, G. E. (1999). Solubilization of phosphates and micronutrients by the plant-growth-promoting and biocontrol fungus *Trichoderma harzianum* Rifai 1295-22. *Applied and Environmental Microbiology, 65(7)*, 2926–2933.

American Phytopathological Society. (2005). Compendium of bean diseases. (Schwartz HF, Steadman J R, Hall R, and Forster R L, Eds.) (2nd ed.). APS Press, American Phytopathological Society.

Atkinson, M. M., Midland, S. L., Sims, J. J. and Keen, N. T. (1996). Syringolide 1 triggers Ca2 influx, K efflux, and extracellular alkalization in soybean cells carrying the disease-resistance gene *Rpg4*. *Plant Physiology, 112(1)*, 297–302. http://doi.org/10.1104/pp.112.1.297.

Bakshi, M. and Oelmüller, R. (2014). WRKY transcription factors. *Plant Signaling & Behavior, 9(2)*, e27700. http://doi.org/10.4161/psb.27700.

Barroso, C. B. and Nahas, E. (2005). The status of soil phosphate fractions and the ability of fungi to dissolve hardly soluble phosphates. *Applied Soil Ecology, 29(1)*, 73–83. http://doi.org/10.1016/j.apsoil.2004.09.005.

Beebe, S. E., Corrales, M. P., Schoonhoven, A. van and Voysest, O. (1991). Breeding for disease resistance. In O. Schoonhoven, A. van;Voysest (Ed.), *Common beans: research for crop improvemen*t. (pp. 561–617). Wallingford: CAB International.

Benítez, T., Rincón, A. M., Limón, M. C. and Codón, A. C. (2004). Biocontrol mechanisms of *Trichoderma* strains. *International Microbiology, 7(4)*, 249–260. http://doi.org/10.2436/IM.V7I4.9480.

Birkenbihl, R. P., Diezel, C. and Somssich, I. E. (2012). *Arabidopsis* WRKY33 is a key transcriptional regulator of hormonal and metabolic responses toward *Botrytis cinerea* infection. *Plant Physiology, 159(1)*, 266–285. http://doi.org/10.1104/pp.111.192641.

Boller, T. and Felix, G. (2009). A Renaissance of elicitors: perception of microbe-associated molecular patterns and danger signals by pattern-recognition receptors. *Annual Review of Plant Biology, 60(1),* 379–406. http://doi.org/10.1146/annurev.arplant.57.032905.105346.

Borges, A., Melotto, M., Tsai, S. M. and Caldas, D. G. G. (2012). Changes in spatial and temporal gene expression during incompatible interaction between common bean and Anthracnose pathogen. *Journal of Plant Physiology, 169(12),* 1216–1220. http://doi.org/10.1016/j.jplph.2012.04.003.

Brimner, T. A. and Boland, G. J. (2003). A review of the non-target effects of fungi used to biologically control plant diseases. *Agriculture, Ecosystems & Environment, 100(1),* 3–16. http://doi.org/http://dx.doi.org/10.1016/S0167-8809(03)00200-7.

Broglie, K., Chet, I., Holliday, M., Cressman, R., Biddle, P., Knowlton, S. and Broglie, R. (1991). Transgenic plants with enhanced resistance to the fungal pathogen *Rhizoctonia solani. Science, 254(5035),* 1194–1197. http://doi.org/10.1126/science.254.5035.1194.

Brotman, Y., Briff, E., Viterbo, A. and Chet, I. (2008). Role of swollenin, an expansin-like protein from *Trichoderma*, in plant root colonization. *Plant Physiology, 147,* 779–789. http://doi.org/10.1104/pp.108.116293.

Brunner, K., Peterbauer, C. K., Mach, R. L., Lorito, M., Zeilinger, S. and Kubicek, C. P. (2003). The Nag1 N-acetylglucosaminidase of *Trichoderma atroviride* is essential for chitinase induction by chitin and of major relevance to biocontrol. *Current Genetics, 43(4),* 289–295.

Campelo, M. P., Cardoza, R. E., Lorenzana, A., Hermosa, M. R., Monte, E., Reinoso, B. and Casquero, P. A. (2010). Biological control of phytopathogenic fungi in bean (*Phaseolus vulgaris*, L.) with *Trichoderma atroviride* and *Trichoderma virens*. In P. Miklas & P. Gepts (Eds.), *Bean Improvement Cooperative Fort Collins* (Vol. 53, pp. 114–115). Colorado, USA.

Cardoza, R. E., Vizcaíno, J. A., Hermosa, M. R., Sousa, S., González, F. J., Llobell, A. and Gutiérrez, S. (2006). Cloning and characterization of the *erg1* gene of *Trichoderma harzianum*: Effect of the erg1 silencing on ergosterol biosynthesis and resistance to terbinafine. *Fungal Genetics and Biology, 43(3),* 164–178. http://doi.org/10.1016/j.fgb.2005.11.002.

Chacón, M. R., Rodríguez Galán, O., Benítez Fernández, C. T., Sousa, S., Rey, M., Llobell González, A. and Delgado Jarana, J. (2007). Microscopic and transcriptome analyses of early colonization of tomato roots by *Trichoderma harzianum. International Microbiology 10(1),* 19–27. http://doi.org/10.2436/20.1501.01.4.

Chagas, L. F. B., De Castro, H. G., Colonia, B. S. O., De Carvalho Filho, M. R., Miller, L. D. O. and Chagas, A. F. J. (2016). Efficiency of *Trichoderma* spp. as a growth promoter of cowpea (*Vigna unguiculata*) and analysis of phosphate solubilization and indole acetic acid synthesis. *Brazilian Journal of Botany, 39(2),* 437–445. http://doi.org/10.1007/s40415-015-0247-6.

Chaman, M. E., Copaja, S. V. and Argandoña, V. H. (2003). Relationships between salicylic acid content, Phenylalanine Ammonia-Lyase (PAL) activity, and resistance of barley to aphid infestation. *Journal of Agricultural and Food Chemistry, 51(8),* 2227–2231. http://doi.org/10.1021/jf020953b.

Chang, Y.-C., Chang, Y.-C., Baker, R., Kleifeld, O. and Chet, I. (1986). Increased growth of plants in the presence of the biological control agent *Trichoderma harzianum. Plant Disease, 70(2),* 145–148.

Chen, W., Provart, N. J., Glazebrook, J., Katagiri, F., Chang, H. S., Eulgem, T. and Zhu, T. (2002). Expression profile matrix of *Arabidopsis* transcription factor genes suggests their putative functions in response to environmental stresses. *The Plant Cell, 14(3)*, 559–574. http://doi.org/10.1105/tpc.010410.

Chen, X., Qi, X. and Duan, L.-X. (2015). Overview. In X. Qi, X. Chen, & Y. Wang (Eds.), *Plant Metabolomics* (pp. 1–24). Dordrecht: Springer Netherlands. http://doi.org/10.1007/978-94-017-9291-2_1.

Chet, I., Harman, G. E. and Baker, R. (1981). *Trichoderma hamatum*: Its hyphal interactions with *Rhizoctonia solani* and *Pythium* spp. *Microbial Ecology, 7(1),* 29–38. http://doi.org/10.1007/BF02010476.

Dana, M. de las M., Pintor-Toro, J. A. and Cubero, B. (2006). Transgenic tobacco plants overexpressing chitinases of fungal origin show enhanced resistance to biotic and abiotic stress agents. *Plant Physiology, 142*, 722–730. http://doi.org/10.1104/pp.106.086140.

Demain, A. L. and Fang, A. (2000). The natural functions of secondary metabolites. In A. Fiechter (Ed.), *History of Modern Biotechnology I* (pp. 1–39). Springer Berlin Heidelberg. http://doi.org/10.1007/3-540-44964-7_1.

Djonović, S., Pozo, M. J., Dangott, L. J., Howell, C. R. and Kenerley, C. M. (2006). Sm1, a proteinaceous elicitor secreted by the biocontrol fungus *Trichoderma virens* induces plant defense responses and systemic resistance. *Molecular Plant-Microbe Interactions, 838(8),* 838–853. http://doi.org/10.1094/MPMI-19-0838.

Druzhinina, I. S., Seidl-Seiboth, V., Herrera-Estrella, A., Horwitz, B. A., Kenerley, C. M., Monte, E. and Kubicek, C. P. (2011). *Trichoderma*: The genomics of opportunistic success. *Nature Reviews Microbiology, 9(10),* 749–759.

Dubey, S. C., Suresh, M. and Singh, B. (2007). Evaluation of *Trichoderma* species against *Fusarium oxysporum* f. sp. *ciceris* for integrated management of chickpea wilt. *Biological Control, 40(1),* 118–127. http://doi.org/10.1016/j.biocontrol.2006.06.006.

Food and Agriculture Organization. (2016). Pulses: Nutritious seeds for a sustainable future. (P. Javaloyes and S. O'Broin, Eds.). *Food and Agriculture Organization of the United Nations*.

Ghisalberti, E. L. and Sivasithamparam, K. (1991). Antifungal antibiotics produced by *Trichoderma* spp. *Soil Biology and Biochemistry, 23(11),* 1011–1020. http://doi.org/10.1016/0038-0717(91)90036-J.

Gravel, V., Antoun, H. and Tweddell, R. J. (2007). Growth stimulation and fruit yield improvement of greenhouse tomato plants by inoculation with *Pseudomonas putida* or *Trichoderma atroviride*: Possible role of indole acetic acid (IAA). *Soil Biology and Biochemistry, 39(8),* 1968–1977. http://doi.org/10.1016/j.soilbio. 2007.02.015.

Grayer, R. J. and Kokubun, T. (2001). Plant–fungal interactions: the search for phytoalexins and other antifungal compounds from higher plants.

Phytochemistry, 56(3), 253–263. http://doi.org/10.1016/S0031-9422(00)00450-7.

Grayston, S. J., Vaughan, D. and Jones, D. (1997). Rhizosphere carbon flow in trees, in comparison with annual plants: the importance of root exudation and its impact on microbial activity and nutrient availability. *Applied Soil Ecology, 5(1),* 29–56. http://doi.org/10.1016/S0929-1393(96)00126-6.

Grotewold, E. (2005). Plant metabolic diversity: a regulatory perspective. *Trends in Plant Science, 10(2),* 57–62. http://doi.org/10.1016/j.tplants.2004.12.009.

Guerrero-González, M. L., Rodríguez-Kessler, M., Rodríguez-Guerra, R., González-Chavira, M., Simpson, J., Sanchez, F. and Jiménez-Bremont, J. F. (2011). Differential expression of *Phaseolus vulgaris* genes induced during the interaction with *Rhizoctonia solani*. *Plant Cell Reports, 30(8),* 1465–1473. http://doi.org/10.1007/s00299-011-1055-5.

Harman, G. E., Howell, C. R., Viterbo, A., Chet, I. and Lorito, M. (2004). *Trichoderma* species--opportunistic, avirulent plant symbionts. *Nature Reviews in Microbiology*, 2(1), 43–56. http://doi.org/10.1038/nrmicro797.

Hermosa, M. R., Grondona, I., Díaz-Mínguez, J. M., Iturriaga, E. A. and Monte, E. (2001). Development of a strain-specific SCAR marker for the detection of *Trichoderma atroviride* 11, a biological control agent against soilborne fungal plant pathogens. *Current Genetics, 38(6),* 343–350. http://doi.org/10.1007/s002940000173.

Hermosa, M. R., Viterbo, A., Chet, I. and Monte, E. (2012). Plant-beneficial effects of *Trichoderma* and of its genes. *Microbiology, 158(1),* 17–25. http://doi.org/10.1099/mic.0.052274-0.

Hermosa, R., Rubio, M. B., Cardoza, R. E., Nicolás, C., Monte, E. and Gutiérrez, S. (2013). The contribution of Trichoderma to balancing the costs of plant growth and defense. International Microbiology, 16(2), 69–80.

Howell, C. R. (2003). Mechanisms employed by *Trichoderma* species in the biological control of plant diseases: the history and evolution of current concepts. *Plant Disease, 87(1)*, 4–10.

Hoyos-Carvajal, L., Orduz, S. and Bissett, J. (2009). Growth stimulation in bean (*Phaseolus vulgaris* L.) by *Trichoderma*. *Biological Control, 51(3),* 409–416. http://doi.org/10.1016/j.biocontrol.2009.07.018.

Huang, F. C., Studart Witkowski, C. and Schwab, W. (2010). Overexpression of hydroperoxide lyase gene in *Nicotiana benthamiana* using a viral vector system. *Plant Biotechnology Journal, 8(7),* 783–795. http://doi.org/10.1111/j.1467-7652.2010.00508.x.

Innocenti, G., Pucciariello, C., Le Gleuher, M., Hopkins, J., de Stefano, M., Delledonne, M. and Frendo, P. (2007). Glutathione synthesis is regulated by nitric oxide in *Medicago truncatula* roots. *Planta, 225(6),* 1597–1602. http://doi.org/10.1007/s00425-006-0461-3.

Kamaruzzaman, M., Rahman, M. M., Islam, M. S., Ahmad, M. U. and Ahmad, M. U. (2016). Efficacy of four selective *Trichoderma* isolates as plant growth promoters in two peanut varieties. *International Journal of Biological Research, 4(2),* 152. http://doi.org/10.14419/ijbr.v4i2.6468.

Kapri, A. and Tewari, L. (2010). Phosphate solubilization potential and phosphatase activity of rhizospheric *Trichoderma* spp. *Brazilian Journal of Microbiology, 41(3),* 787–795. http://doi.org/10.1590/S1517-83822010005000001.

Keszler, A., Forgács, E., Kótai, L., Vizcaíno, J. A., Monte, E. and García-Acha, I. (2000). Separation and identification of volatile components in the fermentation broth of *Trichoderma atroviride* by solid-phase extraction and gas chromatography-mass spectrometry. *Journal of Chromatographic Science, 38(10),* 421–4. http://doi.org/10.1093/CHROMSCI/38.10.421.

Kim, K. C., Fan, B. and Chen, Z. (2006). Pathogen-induced *Arabidopsis* WRKY7 is a transcriptional repressor and enhances plant susceptibility to *Pseudomonas syringae*. *Plant Physiology, 142(3)*, 1180–1192. http://doi.org/10.1104/pp.106.082487.

Lippok, B., Birkenbihl, R. P., Rivory, G., Brümmer, J., Schmelzer, E., Logemann, E. and Somssich, I. E. (2007). Expression of *AtWRKY33* encoding a pathogen-or PAMP-responsive WRKY transcription factor is regulated by a composite DNA motif containing W box elements.

Molecular Plant-Microbe Interactions, 20(4), 420–429. http://doi.org/10.1094/MPMI-20-4-0420.

Lorito, M., Harman, G. E., Hayes, C. K., Broadway, R. M., Tronsmo, A., Woo, S. L. and Di Pietro, A. (1993). Chitinolytic enzymes produced by *Trichoderma harzianum* : antifungal activity of purified endochitinase and chitobiosidase. *Phytopathology, 83(3),* 302–307.

Lorito, M., Hayes, C. K., Di Pietro, A., Woo, S. L. and Harman, G. E. (1994). Purification, characterization, and synergistic activity of a glucan 1,3-β-glucosidase and an N-acetyl-β-glucosaminidase from *Trichoderma harzianum. Phytopathology, 84(4),* 398–405.

Lorito, M., Woo, S. L., Garcia, I., Colucci, G., Harman, G. E., Pintor-Toro, J. A. and Fernandez, I. G. (1998). Genes from mycoparasitic fungi as a source for improving plant resistance to fungal pathogens. *Proceedings of the National Academy of Sciences of the United States of America, 95(14),* 7860–7865.

Lorito, M., Woo, S. L., Harman, G. E. and Monte, E. (2010). Translational research on *Trichoderma* : From *'Omics to the Field. Annual Review of* Phytopathology, 48(1), 395–417. http://doi.org/10.1146/annurev-phyto-073009-114314.

Loscos, J., Matamoros, M. A. and Becana, M. (2008). Ascorbate and homoglutathione metabolism in common bean nodules under stress conditions and during natural senescence. *Plant Physiology, 146(3),* 1282–1292. http://doi.org/10.1104/pp.107.114066.

Lu, Z., Tombolini, R., Woo, S., Zeilinger, S., Lorito, M. and Jansson, J. K. (2004). *In vivo* study of *Trichoderma*-pathogen-plant interactions, using constitutive and inducible green fluorescent protein reporter systems. *Applied and Environmental Microbiology, 70(5),* 3073–81. http://doi.org/10.1128/AEM.70.5.3073-3081.2004.

Malmierca, M. G., Barua, J., Mccormick, S. P., Izquierdo-Bueno, I., Cardoza, R. E., Alexander, N. J. and Gutiérrez, S. (2014). Novel aspinolide production by *Trichoderma arundinaceum* with a potential role in *Botrytis cinerea* antagonistic activity and plant defence priming. *Environmental Microbiology, 17(4),* 1103–1118. http://doi.org/10.1111/1462-2920.12514.

Marra, R., Ambrosino, P., Carbone, V., Vinale, F., Woo, S. L., Ruocco, M. and Lorito, M. (2006). Study of the three-way interaction between *Trichoderma atroviride*, plant and fungal pathogens by using a proteomic approach. *Current Genetics, 50(5),* 307–321. http://doi.org/10.1007/s00294-006-0091-0.

Mauch-Mani, B. and Slusarenko, A. J. (1996). Production of salicylic acid precursors is a major function of Phenylalanine Ammonia-Lyase in the resistance of *Arabidopsis* to *Peronospora parasitica. The Plant Cell, 8(2),* 203–212. http://doi.org/10.1105/tpc.8.2.203.

Mayo, S., Cominelli, E., Sparvoli, F., González-López, O., Rodríguez-González, A., Gutiérrez, S. and Casquero, P. A. (2016). Development of a qPCR strategy to select bean genes involved in plant defense response and regulated by the *Trichoderma velutinum – Rhizoctonia solani* interaction. *Frontiers in Plant Science, 7,* 1109. http://doi.org/10.3389/fpls.2016.01109.

Mayo, S., Gutierrez, S., Malmierca, M. G., Lorenzana, A., Campelo, M. P., Hermosa, R. and Casquero, P. A. (2015). Influence of *Rhizoctonia solani* and *Trichoderma* spp. in growth of bean (*Phaseolus vulgaris* L.) and in the induction of plant defense-related genes. *Frontiers in Plant Science, 6*, 685. http://doi.org/10.3389/fpls.2015.00685.

Mayo, S., Izquierdo, H., González-López, Ó., Rodríguez-González, Á., Lorenzana, A., Gutiérrez, S. and Casquero, P. A. (2016). Effect of farnesol, a compound produced by *Trichoderma* when growing on bean (*Phaseolus vulgaris* L.). In *Joint Natural Products Conference. Copenhague*.

Monte, E. (2001). Understanding *Trichoderma*: between biotechnology and microbial ecology. *International Microbiology, 4*, 1–4. http://doi.org/10.1007/s101230100001.

Montero, M., Sanz, L., Rey, M., Llobell, A. and Monte, E. (2007). Cloning and characterization of bgn16·3, coding for a β-1,6-glucanase expressed during *Trichoderma harzianum* mycoparasitism. *Journal of Applied Microbiology, 103(4),* 1291–1300. http://doi.org/10.1111/j.1365-2672.2007.03371.x.

Montero, M., Sanz, L., Rey, M., Monte, E. and Llobell, A. (2005). BGN16.3, a novel acidic β-1,6-glucanase from mycoparasitic fungus *Trichoderma harzianum* CECT 2413. *FEBS Journal, 272(13),* 3441–3448. http://doi.org/10.1111/j.1742-4658.2005.04762.x.

Noordermeer, M. A., Veldink, G. A. and Vliegenthart, J. F. G. (2001). Fatty acid hydroperoxide lyase: a plant cytochrome P450 enzyme involved in wound healing and pest resistance. *Chembiochem, 2(7-8),* 494–504. http://doi.org/10.1002/1439-7633(20010803)2:7.

Nugroho, L. H., Verberne, M. C. and Verpoorte, R. (2002). Activities of enzymes involved in the phenylpropanoid pathway in constitutively salicylic acid-producing tobacco plants. *Plant Physiology and Biochemistry, 40(9),* 755–760. http://doi.org/10.1016/S0981-9428(02)01437-7.

Öğüt, M. and Er, F. (2006). Micronutrient composition of field-grown dry bean and wheat inoculated with *Azospirillum* and *Trichoderma. Journal of Plant Nutrition and Soil Science, 169(5),* 699–703. http://doi.org/10.1002/jpln.200520597.

Pandey, S. P. and Somssich, I. E. (2009). The role of WRKY transcription factors in plant immunity. *Plant Physiology, 150(4)*, 1648–1655. http://doi.org/10.1104/pp.109.138990.

Paulitz, T. C. (1990). Biochemical and ecological aspect of competition in biological control. In R. R. Baker & P. E. Dunn (Eds.), *New directions in biological control. Alternatives for suppressing agricultural pests and diseases* (p. 837). New York (USA): Wiley-Liss Inc.

Pereira, J. L., Queiroz, R. M. L., Charneau, S. O., Felix, C. R., Ricart, C. A. O., Lopes Da Silva, F. and Noronha, E. F. (2014). Analysis of *Phaseolus vulgaris* response to its association with *Trichoderma harzianum* (ALL-42) in the presence or absence of the phytopathogenic fungi *Rhizoctonia solani* and *Fusarium solani. PLoS ONE, 9(5),* e98234. http://doi.org/10.1371/journal.pone.0098234.

Reino, J. L., Guerrero, R. F., Hernández-Galán, R. and Collado, I. G. (2007). Secondary metabolites from species of the biocontrol agent *Trichoderma. Phytochemistry Reviews, 7(1),* 89–123. http://doi.org /10.1007/s11101-006-9032-2.

Rojo, F. G., Reynoso, M. M., Ferez, M., Chulze, S. N. and Torres, A. M. (2007). Biological control by *Trichoderma* species of *Fusarium solani* causing peanut brown root rot under field conditions. *Crop Protection, 26(4),* 549–555. http://doi.org/10.1016/j.cropro.2006.05.006.

Rosado, I. V., Rey, M., Codón, A. C., Govantes, J., Moreno-Mateos, M. A. and Benítez, T. (2007). QID74 Cell wall protein of *Trichoderma harzianum* is involved in cell protection and adherence to hydrophobic surfaces. *Fungal Genetics and Biology, 44(10),* 950–964. http://doi.org/10.1016/j.fgb.2007.01.001.

Rubio, M. B., Hermosa, M. R., Keck, E. and Monte, E. (2005). Specific PCR assays for the detection and quantification of DNA from the biocontrol strain *Trichoderma harzianum* 2413 in soil. *Microbial Ecology, 49(1),* 25–33. http://doi.org/10.1007/s00248-003-0171-3.

Ruocco, M., Lanzuise, S., Vinale, F., Marra, R., Turrà, D., Woo, S. L. and Benítez, T. (2007). Identification of a new biocontrol gene in *Trichoderma atroviride*: the role of an ABC transporter membrane pump in the interaction with different plant-pathogenic fungi. *Molecular Plant-Microbe Interactions, 44(3)*, 950–964. http://doi.org /10.1016/j.fgb.2007.01.001.

Rushton, P. J. and Somssich, I. E. (1998). Transcriptional control of plant genes responsive to pathogens. *Current Opinion in Plant Biology, 1(4),* 311–315. http://doi.org/10.1016/1369-5266(88)80052-9.

Sanz, L., Montero, M., Grondona, I., Vizcaíno, J. A., Llobell, A., Hermosa, R. and Monte, E. (2004). Cell wall-degrading isoenzyme profiles of *Trichoderma* biocontrol strains show correlation with rDNA taxonomic species. *Current Genetics, 46(5)*, 277–286. http://doi.org/ 10.1007/s00294-004-0532-6.

Sanz, L., Montero, M., Redondo, J., Llobell, A. and Monte, E. (2005). Expression of an α-1,3-glucanase during mycoparasitic interaction of *Trichoderma asperellum. FEBS Journal, 272(2),* 493–499. http://doi.org/10.1111/j.1742-4658.2004.04491.x.

Seidl, V., Marchetti, M., Schandl, R., Allmaier, G. and Kubicek, C. P. (2006). Epl1, the major secreted protein of *Hypocrea atroviridis* on glucose, is a member of a strongly conserved protein family comprising plant defense response elicitors. *FEBS Journal, 273(18),* 4346–4359. http://doi.org/10.1111/j.1742-4658.2006.05435.x.

Shoresh, M., Harman, G. E. and Mastouri, F. (2010). Induced systemic resistance and plant responses to fungal biocontrol agents. *Annual Review of Phytopathology, 48,* 21–43. http://doi.org/10.1146/annurev-phyto-073009-114450.

Singh, K. B., Foley, R. C. and Oñate-Sánchez, L. (2002). Transcription factors in plant defense and stress responses. *Current Opinion in Plant Biology, 5(5),* 430–436. http://doi.org/10.1016/S1369-5266(02)00289-3.

Sivasithamparam, K. and Ghisalberti, E. L. (1998). Secondary metabolism in *Trichoderma* and *Gliocladium*. In C. P. Kubicek & G. E. Harman (Eds.), *Trichoderma and Gliocladium: Basic biology, taxonomy and genetics* (pp. 139–191). London, United Kingdom: Taylor and Francis Ltd.

Suarez, B., Rey, M., Castillo, P., Monte, E. and Llobell, A. (2004). Isolation and characterization of PRA1, a trypsin-like protease from the biocontrol agent *Trichoderma harzianum* CECT 2413 displaying nematicidal activity. *Applied Microbiology and Biotechnology, 65(1),* 46–55. http://doi.org/10.1007/s00253-004-1610-x.

Valenciano, J. B., Casquero, P. A. and Boto, J. A. (2004). Influence of sowing techniques and pesticide application on the emergence and the establishment of bean plants (*Phaseolus vulgaris* L.). *Agronomie, 24(2),* 113–118. http://doi.org/10.1051/agro:2004007.

Veitch, N. C. (2009). Isoflavonoids of the Leguminosae. *Natural Product Reports, 26(6),* 776. http://doi.org/10.1039/b616809b.

Vinale, F., Sivasithamparam, K., Ghisalberti, E. L., Marra, R., Woo, S. L. and Lorito, M. (2008). *Trichoderma*–plant–pathogen interactions. *Soil Biology and Biochemistry, 40(1),* 1–10. http://doi.org/10.1016/j.soilbio.2007.07.002.

Viterbo, A., Montero, M., Ramot, O., Friesem, D., Monte, E., Llobell, A. and Chet, I. (2002). Expression regulation of the endochitinase chit36 from *Trichoderma asperellum (T. harzianum* T-203). *Current Genetics, 42(2),* 114–122. http://doi.org/10.1007/s00294-002-0345-4.

Viterbo, A., Wiest, A., Brotman, Y., Chet, I. and Kenerley, C. (2007). The 18mer peptaibols from *Trichoderma virens* elicit plant defence responses. *Molecular Plant Pathology, 8(6)*, 737–746. http://doi.org/10.1111/j.1364-3703.2007.00430.x.

Vizcaíno, J. A., Cardoza, R. E., Hauser, M., Hermosa, R., Rey, M., Llobell, A. and Monte, E. (2006). ThPTR2, a di/tri-peptide transporter gene from *Trichoderma harzianum. Fungal Genetics and Biology, 43(4),* 234–246. http://doi.org/10.1016/j.fgb.2005.12.003.

Vizcaino, J. A., Sanz, L., Basilio, A., Vicente, F., Gutierrez, S., Hermosa, M. R. and Monte, E. (2005). Screening of antimicrobial activities in *Trichoderma* isolates representing three *Trichoderma* sections. *Mycological Research, 109(12),* 1397–1406. http://doi.org/10.1017/S0953756205003898.

Woo, S. L. and Lorito, M. (2007). Exploiting the interactions between fungal antagonists, pathogens and the plant for biocontrol. In M. Vurro & J. Gressel (Eds.), *Novel Biotechnologies for Biocontrol Agent Enhancement and Management* (pp. 107–130). The Netherlands: Springer.

Woo, S. L., Scala, F., Ruocco, M. and Lorito, M. (2006). The molecular biology of the interactions between *Trichoderma* spp., phytopathogenic fungi, and plants. *Phytopathology, 96(2),* 181–185.

Yedidia, I., Srivastva, A. K., Kapulnik, Y. and Chet, I. (2001). Effect of *Trichoderma harzianum* on microelement concentrations and increased growth of cucumber plants. *Plant and Soil, 235(2),* 235–242.

In: Legumes for Global Food Security ISBN: 978-1-53612-265-7
Editors: Alfonso Clemente et al.

Chapter 5

PARTICIPATORY ACTION RESEARCH METHODOLOGIES TO ADVANCE LEGUME PRODUCTIVITY

***Simon Mark Payne*[1,*], *Phillipa K. Nicholas-Davies*[1] and *Robert Home*[2]**
[1]Institute of Biological, Environmental and Rural Sciences, Aberystwyth University, Wales
[2]Department of Socio-Economic Science, Research Institute of Organic Agriculture, Switzerland

ABSTRACT

The aim of this chapter is to provide readers with a rationale for, and practical guide to beginning or expanding their repertoire of participatory research techniques. The chapter merges insight from cognitive psychology and social psychology with evidence of good practice in participatory action research from agricultural science. It is hoped that readers will be inspired and better equipped to foreground the individuals

* Corresponding Author Email: smp14@aber.ac.uk.

involved in their research over the "hard science"; this is a trusted path to research *impact*, the importance of which has never been greater.

Keywords: social cognition, social learning, empowerment, collaboration, managing change, psychology, individual differences

1. Introduction

The United Nations (UN) recently announced the 17 goals that spearhead their 2030 agenda for sustainable development. The second goal on this ambitious list – "Zero Hunger" – aims to: "End hunger, achieve food security and improved nutrition and promote sustainable agriculture" (UN, 2015). Recent research has identified legumes as an essential food crop and a vital source of plant-based protein in this fight (Abraham et al., 2014; Courty et al., 2014; de Bruijn, 2015; Duranti & Scarafoni, 2015; see also, 2015 "Special Issue on Legumes in Sustainable Agriculture," *Critical Reviews in Plant Sciences*, 34), but wherever new techniques are introduced and intensified production is expected (i.e., *change*), the human element is foregrounded; no matter the experimental evidence supporting certain production techniques, farmers must be willing and able to adopt and sustain the recommended changes. Indeed, farmers' decision making processes are complex (Barlett, 1980; Darnhofer, Schneeberger, & Freyer, 2005; Öhlmér, Olson, & Brehmer, 1998; Struik, Kuyper, Brussaard, & Leeuwis, 2014); their ability to maintain altered farming behaviours are dependent on myriad factors, for example, personality, cognitive ability, and coping skills, and not least their attitudes and goals (Willock et al., 1999). Moreover, farmer decisions and actions are influenced not just by their own psychology and socio-demographics, characteristics of the farm household, structure of the farm business, and characteristics of the innovation to be adopted (Edwards-Jones, 2006, p. 783), but by other actors in the value chain (input suppliers, processors, retailers, consumers) and the communities in which they farm (Meijer et al., 2015; Tilman, Cassman, Matson, Naylor, & Polasky, 2002). Hence, research that explores the feasibility of advanced legume production will enhance its potential impact by building into its design, research strategies that enlist such actors as co-investigators of knowledge.

Agricultural and food scientists make recommendations for *change* – evidence-based suggestions that they believe will improve the situation we find ourselves in. For example, the US National Research Council's (2010) *Toward Sustainable Agricultural Systems in the 21st Century*, and *Achieving food security in the face of climate change*, a report from the Commission on Sustainable Agriculture and Climate Change (Consultative Group on International Agricultural Research, 2012). The most robust recommendations acknowledge that there are as many barriers to change as there are individuals within a value chain (and organisational forces exerted on these individual actors). In the Australian farming context, Guerin and Guerin (1994) identified major constraints to the adoption of technologies and innovations, including those which go beyond farmer education issues that are traditionally addressed by agricultural extension (e.g., "the farmer's beliefs and opinions towards the technology; the farmer's level of motivation; the farmer's perception of the relevance of the new technology; and the farmer's attitudes towards risk and change"; p. 549). In terms of risk, Liu (2013) found that Chinese farmers who were more risk averse or more loss averse were slower to adopt genetically modified cotton and thus accumulated less wealth. Liu concluded that, as per her hypotheses, fear of loss and aversion to risk *do* deter technological advancement, and inefficient input allocation is caused by risk in agricultural production.

As a final example, Asfaw, Shiferaw, Simtowe, and Lipper (2012) investigated barriers to adoption of improved legume varieties in smallholdings in Tanzania and Ethiopia. They uncovered many determinants of technology adoption – from distance to nearest agricultural office, contact with extension agent, number of improved varieties known, to perceptions of these varieties – highlighting the complex nature of farmer decision making. Hence, despite some universal barriers to adoption, each crop and each geographical region will pose unique challenges. Moreover, *each personality involved* in the supply chain will influence the dynamics of the wider system. The key, therefore, is to be armed with a research methodology that allows for – indeed, welcomes – such idiosyncrasies and diversity. Knowledge of the barriers that motivated farmers must overcome can be teased out with participatory research strategies, particularly in the early stages, with their participation in “needs and opportunity assessment, setting of research and development objectives, and establishment of indicators” (van de Fliert & Braun, 2002, p. 26). This suggestion is set against the backdrop of research funding at the applied end of the scale being increasingly evaluated in terms of its potential impact rather than just academic excellence.

Thus, the primary aim of this chapter is to provide hands-on examples of participatory research techniques that can be used by researchers who want to maximise the impact of their legumes research to deliver outcomes such as, for example, improved soil quality and food security, and other economic and environmental benefits.

2. What Is Participatory Action Research?

The *participatory action research* (PAR) approach shares some common philosophical underpinnings with, and thus blends elements of *participatory* research and *action* research methodologies. For example, PAR breaks down the distinction between ‘researcher’ and ‘participant’ by considering each as participants in a process of co-creation of new knowledge (Home & Rump, 2015). PAR invites prospective beneficiaries of the research to bring their expertise to bear on some or all of four key tasks: defining the problem, choosing the methods, analysing the data, and putting the findings to best use (Elden, Reason, & Rowan, 1981). After the participants’ and researchers’ (read, *collaborators*) perspectives are jointly incorporated into the research design and the problem is defined, PAR typically tasks both parties with continuing to work on the problem independently *and* in tandem, as a three-pronged quality assurance check of sorts (Grundy, 1987). PAR collaborators/collaborations are reflective, and reflection is built in to the process and occurs in real time, meaning that change is enveloped into the ongoing research and its necessity is not evaluated *post hoc* (Chevalier & Buckles, 2013). Thus, where necessary, theoretical and practical standpoints can be reconciled through action (Whyte, 1989).

PAR is an umbrella term representing a constellation of research strategies and techniques which share the same goals and guiding principles, can be selected to suit the context at hand, and flexibly employed to suit the precise aims of each project. Engaging in PAR is motivated by a particular philosophical stance, i.e., one's beliefs regarding the nature of reality and the means by which knowledge can be created (epistemology and ontology). Specifically, PAR might be underpinned by a constructivist ontology – which views 'reality' as subjective, multiple, and constructed through the interaction of the researcher and researched (Charmaz, 2006) – and interpretivist epistemology, which posits that: "direct knowledge of phenomena is not possible…observations and accounts of the world provide indirect indications of phenomena, and thus knowledge is developed through a process of interpretation" (Weed, 2009, p. 507). Constructivist-interpretivist research stands in contrast to positivism, which is based on a realist ontology, reflecting: "the belief that there exists a [singular] 'reality' *out there*, driven by immutable laws of nature" (Guba, 1990, p. 19; parentheses added) – and the researcher's approach is deductive, their role purely one of measurement. In the experience of the present authors and in the experience of colleagues with which they have discussed this issue, it seems to be this philosophy that puts off some scientists from making their research more participatory – they don't share the same view of reality and how knowledge is created; their view is narrower (positivist) and perhaps not so well attuned to the needs of those in "contexts where uncertainties are high, and problems are open to interpretation" (Pretty, 1995).

Hence, legume-related PAR guided by these complementary constructivist-interpretivist philosophies accepts that *reality* means something different for agricultural, industrial, and social partners representing each step of the value chain (input suppliers, farmers, processors, retailers, consumers) and, indeed, the physical and social scientists with which they collaborate. Scientists who collect data on farmers' intentions to adopt their recommendations and resultant crop yield, for example, only uncover a very narrow portion of the 'story'; it is prudent instead for the scientists and participants – co-creators of new, useful knowledge – to interactively interpret each other's *observations and accounts* as a window on the *world* as it is lived.

3. What Is Social Cognition and Social Learning?

With a rich, impactful history in diverse domains of human existence, social cognitive theory (SCT; Bandura, 1986) explains the developmental changes that people experience during their life-course. Specifically, SCT posits that human motivation and behaviour involve *triadic reciprocal determinism*, that is, "behavior (sic), cognition and other personal factors [e.g., conceptions, beliefs, self-perceptions, aspirations, and intentions], and environmental influences all operate as interacting determinants that influence each other bidirectionally" (Bandura, 1989, p. 2). The relative strength of each influence to motivation and behaviour is unique to each situation: it depends of the focus of the activity, circumstances within the immediate situation, and constraints and opportunities provided by the prevailing sociostructural environment. In the context of sustainable legume production, therefore, it is important to understand that the decisions of stakeholders – manifest as behaviours – are a result of, *and contribute to* future interactions between, their thoughts, moods, emotions, biographies, and prevailing physical, geographical and cultural milieu. For example:

> What does the farmer *think* about a crop or technology? What has she/he *heard* about it? How would adopting it fit with her/his *goals* for the upcoming season? Does the farmer have a history of avoiding change? Does the new crop provide the subsistence farmer with food that suits her/his family's tastes? Would adopting it detract from an alternative long-standing preference? Does the farmer have access to

appropriate information and know who to ask for impartial advice? Is *that* person armed with the requisite knowledge? What mind-set does the farmer bring with them on the day of a focus group interview or dissemination event? How does the farmer perceive their role in a participatory research project? Has their role been communicated accurately and the information received as intended? Does it motivate the farmer to engage? Do they feel that their voice has been – and will continue to be – heard? In what mind-set does the farmer *return* to their farm? What day-to-day realities are they facing on the farm that might distract them from continuing their active involvement in a project? What barriers will they face that will detract from their *intention* to change?

Researchers whose work has brought them into direct contact with stakeholders in the value chain will be able to add many more examples to this list.

If, as repeatedly predicted by SCT, (1) thoughts effect behaviours, (2) environmental influences alter thoughts, (3) behaviours influence the environment, and the reverse is also true of each, then powerful insight can be gained by "opening a window" on the world of one's stakeholders – 'inside information' that goes a long way to establishing the conditions under which farmers and other stakeholders are able to, or not, implement the recommendations made by scientists. Hence, if researchers can discover these truths during the design phase of an ostensibly experiment-based legume project, they will maximise the likely achievement of research *impact*. The best method for acquiring this information is to *ask* and involve these actors in the research process. Positivists might believe that scientific *fact*, 'is fact,' and expect this to be enough to convince farmers to act on their recommendations (Pretty, 1995); post-positivists, or interpretivists, on the other hand, with their divergent philosophies of knowledge and knowledge acquisition, are more amenable to scientist-participant collaboration (PAR!).

Ultimately, for new legume production techniques to be successfully promoted farmers should participate in PAR projects that help them overcome any barriers that they have and turn intentions into action. Psychological factors which mediate farmers' knowledge-intention-action process include: (a) the salience of their motives for change and the strength of their motivation to act on these motives; (b) the strength versus susceptibility-to-change of their prevailing habits; (c) their appraisal of the stress of change and risk and their associated coping strategies; (d) their self-efficacy for change itself as well as using the newly recommended agricultural techniques; (e) the extent and nature of their social support network; (f) group dynamics within their immediate social contexts and within their supply chain; and (g) their individual learning styles and preferences. It is therefore imperative to explore these topics with farmers in the early stages of PAR. The social element in all of these areas is apparent – farmers' motivation, prevailing habits, stressor appraisals, coping resources, self-efficacy for change, etc., are not static constructs and they are all modifiable – to varying extents – via social interaction and social learning.

In the case of agriculture research, participatory methods are empowering: giving farmers and members of the supply chain a voice engenders "buy in" that can help overcome a disconnect that they often experience with the science (Snapp, Blackie, & Donovan, 2003; Lyon, Bell, Gratton, & Jackson, 2011; Kirchhoff, Lemos, & Dessai, 2013; cf. Ingram et al., 2016;). Beyond having a voice and being confident that it is heard, participatory methods can help collaborators learn to appreciate and harness their status *as* a knowledge system *within* a knowledge system (Leeuwis & Pyburn, 2002). The actors in legume production do not act independently, they are members of a *system* and they learn from and through participation in such systems (Horton & Mackay, 2003). However, they may also be motivated to maintain autonomy within the system (Bellows & Hamm, 2001), and again, this is an example of where PAR methods can break down barriers to optimal individual and group functioning.

Learning is important in social cognitive theory; stakeholders in legume production may learn about the practical demands of their task from trial and error, yes, but learning can be accelerated through observation of pertinent models (e.g., other legume farmers) – both directly via instruction that is solicited or volunteered, and indirectly with the sharing of good practice, aided by open communication channels between actors (cf. Bandura, 1989). Hence, scientists should similarly learn from their non-science partners to enhance their own capacity to understand challenges faced 'on the ground.' Furthermore, if they are coordinating research that involves stakeholders who would otherwise not come into contact or perhaps be non-communicative, they can facilitate the social learning process with effective and regular dissemination of learning points (act as intermediaries or facilitators). A comprehensive literature search suggests that relatively speaking, most social learning theory-inspired research on the natural environment seems to have been conducted in water management, and the reader is directed to Muro and Jeffrey (2008) for a review.

4. Participatory Action Research Methodologies to Advance Legume Productivity

On reviewing the literature it becomes immediately apparent that the relative degree of stakeholder engagement varies widely in legumes research which claims to be *participatory* (PR). Indeed, much legumes research is participatory but does not quite fit the description of participatory action research (PAR) given earlier in the chapter. In fact, there is lively debate over how best to "blend various forms and intensities of stakeholder participation with quality agricultural science," of which Neef and Neubert (2010, p. 179) provide an excellent summary. Their resultant framework forwards six dimensions of participatory research – project type, research approach, researchers' characteristics, researcher-stakeholder interaction, stakeholders' characteristics, and stakeholders' benefits – and can be seen as a non-legume-specific complement to the present formulation (Neef & Neubert, 2010). Similarly, there is considerable scope regarding the extent to which authors describe the philosophical underpinnings of their participatory research, and whether or not psychological elements – such as those foregrounded by social cognitive theory and social learning theory – are explicitly targeted.

In any case, it is clear that PR and PAR is rarely used in developed countries to increase the use and improve the management of legumes to deliver environmental benefits. In Europe, particularly, the mechanism to encourage behavioural change with respect to legume use, has been through farmer support payments to encourage them to grow legumes crops. The reformed CAP and "greening" have sustained this incentivisation approach, but the impact of these policies on the area of legumes grown and farmer behavioural changes with respect to the use of legumes has not been measured. A review of seven current or recently completed EU-funded projects related to legumes (including Protein2Food, LegumePlus, Healthy Hay, Legume Futures, Legato, Eurolegume and Grain Legumes) found that they had a strong science emphasis, with a certain level of stakeholder engagement (e.g., stakeholder advisory groups, dissemination activity) but with little or no engagement with end-users in the *actual* research process. Projects related to mixed farming and organic farming, such as the recently completed Cantogether and Sustainable Organic and Low Input Dairying (SOLID) projects are two of the relatively few studies that have included participatory research related to legume use and management, albeit indirectly through the role of legumes in the mixed and organic farming systems themselves. In the SOLID project farmers identified and prioritised research needs (in workshop discussions) in each participating country, then a subset of farmers worked with researchers to develop a research methodology and conduct the research on their farms or with a group of farmers.

On the other hand, PR, and to a lesser extent PAR, are used *widely* in developing countries/regions (e.g., Malawi and sub-saharan Africa) to increase the use and improve the management of legumes to achieve a wide range of farmer and community aims and objectives. The objectives of this research fall into several different themes including soil fertility building, identifying suitable plant varieties, sustainability of natural resources and the farm/farming family, resilience and increased food security. Examples of participatory research undertaken in these themes and which are related to the use of legumes are presented and discussed in the next section of this chapter.

A recent report from the European Innovation Partnership (EIP) Focus Group on Protein Crops (2014) highlighted the need to transition from the present level of growing and use of legumes in Europe to increased European plant protein self-sufficiency. The Group believe that this will require not just research by scientists but increased co-operation throughout the legume crop value chain and strong links and communication between stakeholders, including farmers, industry and researchers. The Group "agreed that a coherent research programme should ensure protein crop innovations in practice rather than ensuring research" (p. 14). Hence, legume advances that have been achieved with PR and PAR in developing countries should be taken as proof positive for increased use of PR and PAR in European legumes research.

5. Does PR Work?

In a review of participatory plant breeding (PBB), Ashby and Lilja (2004) showed how participatory research increases benefits in terms of (i) improving research efficiency (by identifying acceptable varieties relatively quickly and therefore accelerating the rate of uptake) and (ii) more effectively reaching women and the poor. Poor farmers are often trapped in situations where they are degrading their natural resources and lack access to more productive and sustainable technologies. One reason for this is that available technology is often unsuitable for them given their objectives and constraints. Hence, empowering these farmers to influence the kinds of technologies that are being developed has the potential to enhance the contribution of agricultural research to productivity, sustainability, and, ultimately, poverty alleviation (Johnson et al., 2003). The involvement of women in maize PPB in China resulted in a broadened national maize genetic base, improved maize yields and strengthened women's' organizations (Song, 1998). Ashby and Lilja (2004) also found that in order to achieve the positive impacts of PR in plant breeding described above, the careful choice of research goals, targeting of environments and selection of user communities is necessary. Additionally, understanding the type of participation required and selecting appropriate techniques and tools to implementation is essential.

Nederlof and Dangbégnon (2007) identified five key criteria that should be used when developing PR programmes with resource-poor farmers, however, this could be equally applied to PR with all farmers. The criteria include:

1. The researcher/research commissioning body has thorough understanding of the context and relevant stakeholders before the start of the project as this will allow identification of existing opportunities or potential for innovation at a macro level.
2. Research is grounded in the opportunities, needs and expectations of intended beneficiaries and research activities are anchored in local conditions.
3. Research results in innovations and technologies that work under the conditions facing farmers.
4. Research results are acceptable and appropriate, implying that the innovation or technology not only yields the desired results but also fits the culture, preferences, traditions, personal circumstances and priorities of the end users.
5. Research develops innovations and technologies that have an impact beyond that of the farmers initially involved and past the duration of the funded project.

When used appropriately and in an expert manner, PR is a proven complement to conventional non-participatory research approaches and is increasingly recognized and practised as an integral part of modern innovation systems. Non-experts should not be disheartened, however, as the remainder of the chapter serves as a practical guide to conducting psychologically-informed PR/PAR in legumes. The boxes presented amongst the text below describe a number of PR projects in the area of legumes research including the aims of the research, what techniques were used in the research process, what worked and what did not and what lessons can be learned from the projects. The lessons learned from these studies in many cases map onto the key criteria for PR research identified by Nederlof and Dangbégnon (2007).

6. Legumes and Participatory Research

Soil Fertility Building

Soil erosion and soil fertility loss are considered to be undermining the productive capacity of sub-Saharan Africa soils. Smallholder farmers use a wide range of agro-ecological management techniques, resource management practices and production strategies specific to their environment to minimize risk, cope with change and shocks, and to manage the environment they operate within (ecological, social, and economic; Nyende & Delve, 2004). These strategies can include, for example, agricultural intensification, expanded market-orientation, and increased capital and labour investment. However, when financial and labour constraints are too prohibitive, or environmental conditions too variable for secure investment, the soil resource base can end up being depleted. For example, a number of studies (Defoer et al., 1998; Ghassali et al., 1999; Witcombe, 1999) have suggested that only relatively well-off farmers have an interest in, and sufficient resources, to prioritise soil improvement criteria for selecting crop varieties and cropping systems. During challenging times it is beneficial to recognise that the farmer's decision making will reflect an interplay between biographical, psychological, social and environmental determinants, and non-participatory research, however excellent, will miss vital cues that could lead to *impact*.

Table 1. Example of participatory research to evaluate the soil fertility building properties of legume crops

Authors and study area	Nyende and Delve (2004), Uganda
Legume study theme	Soil fertility
Study Aim	Farmer participatory evaluation of legume crops for soil fertility building.
Participatory process	Farmer focus groups to identify preference ranking criteria; Farmer field evaluation of the legume crops; Follow up farmer focus groups to evaluate results and identify further research questions.
Aspects of participatory study success	Participatory process was essential to ground the researchers crop recommendations in the reality of farming and the complex range of objectives that farmers may have from a crop, in addition to fertility building, that determine its acceptability and usefulness.
Constraints to participatory study success	Only relatively well-off farmers have an interest in, and sufficient resources, to prioritise soil improvement criteria for selecting crop varieties and cropping systems. In many instances, other complex criteria such as food security and labour take priority.
Lessons to learn	Frame research questions in the context of participants priorities (i.e., not necessarily soil fertility in this case) in order to engage them in the process (Criteria 2).

Nyende and Delve (2004) used PR methods in Uganda to evaluate a range of legume species for building soil fertility. Farmers had multiple criteria for assessing and selecting those species that fitted within their production systems and production objectives. Farmers also adapted the technologies to allow for local opportunities and constraints. Of the five legume species identified by researchers as being suitable for fertility building in the environmental conditions, only one was considered acceptable by the farmers. The participatory process was essential in this instance to ground the researchers' recommendations in the reality of farming and the complex range of objectives that farmers may have from a crop, in addition to fertility building, that determine its acceptability and usefulness. Another two studies (Carberry et al., 2004; Defoer et al., 1998) used PAR methods combined with modelling to engage small holder farmers in Zimbabwe and Mali, respectively, on issues related to soil fertility management. Both studies included the use of legumes with crop rotations to improve soil fertility and yield, alongside other management strategies. Both studies found that combining the system modelling (running different scenarios of how management changes could impact on the farm) with farmer discussion and feedback, provided a good framework for engagement and subsequent development of participatory, on-farm experimentation aimed specifically at the needs of the farmer. Clearly, these are strong examples of research that harnesses the *triadic reciprocal determinism* between farmer behaviour (direct, active involvement), cognition and other personal factors (teased out through discussion), and environmental influences (via modelling), to its benefit.

Participatory Plant Breeding

Participatory plant breeding (PPB) has been widely used in developing countries to select suitable varieties for further development. PPB is a set of approaches that apply in situations where client demand for different varietal traits is poorly understood and difficult to diagnose with conventional market research methods. When choosing a cultivar farmers consider numerous characteristics including yield, marketability, colour, size, growth habit, taste, resistance to pests and diseases, drought tolerance and storability amongst others. Understanding the crop production problems and research priorities of local farmers in various agro-ecological and socio-economic contexts is important to the selection of varieties that will have long term adoption by farmers (Mekbib, 1997). In terms of PPB, researchers can choose to pay lip-service to PAR methods and risk missing distinct pathways to impact, or instead learn to harness the complexity of the farmers' realities.

Table 2. Example of participatory research in legume crop breeding programmes

Authors and study area	Kitch et al., (1998), Northern Cameroon
Legume study theme	Plant breeding
Study Aim	Using participatory research methods to identify farmer selection criteria for cowpea, which then feed into plant breeding programmes.
Participatory process	Breeding trials were conducted on research stations and evaluated by “expert” farmers. Criteria used in selecting experts included ability to communicate, recognition in the community as an expert, several years cultivating cowpea and innovativeness. Individuals were selected to represent a diversity of age, gender, ethnic background and socio-economic status.
Aspects of participatory study success	Farmer evaluations suggest that above a certain minimum acceptable yield, the acceptability of a line is determined by factors other than yield. Plant breeder investment in identifying highest-yielding lines without a corresponding effort to understand farmer preferences would not be cost effective.
Constraints to participatory study success	Broader engagement with the farming family in such studies, rather than just perceived “experts” might highlight other selection criteria, e.g., food security. Research station growing conditions might not be representative of farming.
Lessons to learn	In selecting participants/regions for involvement in a study, consideration should be made of gender, cultural and agro-ecological differences and the impact they might have on results (Criteria 1, 2 and 4). The innovations being tested and the method of testing must reflect the conditions being faced by farmers (Criteria 3).

In a study by Kitch et al., (1998), cowpea (Vigna unguiculata) farmers from diverse geographical areas in northern Cameroon selected breeding lines from on-station trials for their own fields and explained their reasons (criteria) for making the selections. The average selection intensity employed by farmers varied from 6 to 17% and was similar to that employed by plant breeders. Their selection criteria fell into three broad categories relating to yield, preference and quality, and labour, which were very consistent over years, locations and gender. Selection criteria were strongly influenced by market preferences, reflecting the increasing role of cowpea as a cash crop and the prominence of socioeconomic variables in farmer decision making. Ashby and Lilja (2004) emphasise the increasing importance of PPB in a world where there is less public-sector funding and an increasing role in developing country agriculture of supply-driven, non-governmental organizations with little formal accountability to poor farmers. Beyond being a way for researchers to gain feedback from farmers, PPB-as-PAR is a mechanism for farmers, advisors and researchers to form strong collaborative links to develop competitive, future-proof farming systems.

Sustainability

A series of experiments were conducted on communally-owned grasslands in the barley-livestock zone of north Syria. These experiments tested the hypothesis that the introduction of Mediterranean annual legumes would increase productivity and rehabilitate degraded grassland and allow villagers to continue to farm the land (Ghassali et al., 1999). The experiments were preceded by a survey to determine farmers' attitudes, describe the farming systems and to select appropriate collaborators for the field experimentation. The first experiment examined the establishment of medics (Medicago spp.) and clovers (Trifolium spp.) distributed by hand, and monitored their effects on biomass and seedbank size. Later experiments extended these results through participatory methods to other villages. In this research, PR overcame the constraints imposed on the management of this land by its communal ownership, providing solutions that allowed farming the land to become more sustainable.

Table 3. Example of participatory research to determine the role of legumes in the sustainability of maize production

Authors and study area	Mhango et al., (2012), Malawi
Legume study theme	Sustainability
Study Aim	Participatory research to identify opportunities and constraints to legume use for supporting sustainable maize production.
Participatory process	Survey of farmers in project area (Soils, Food and Healthy Communities [SFHC] project) that has provided nutritional and agronomic education to promote legume diversification, and supported gains in soil fertility and family nutrition.
Aspects of participatory study success	Nutrition education, new legume species, and on-farm experimentation through participatory research and extension have increased legume adoption and residue incorporation in the communities studied.
Constraints to participatory study success	Farmers were interested in growing legumes but there were barriers to wider adoption and scaling up, including seed access, labour and markets.
Lessons to learn	Socio-economic, rather than technical constraints are often the key barriers to behaviour change and therefore must be considered by researchers (Criteria 4).

Mhango et al., (2012) sought to identify the opportunities and constraints to legume diversification to support sustainable maize production on smallholder farms in Malawi. The study found that while legumes offer an important means of improving food security and sustainable maize cropping, there were a number of barriers to the adoption of legumes, including limited access to seeds (because of their expense) and labour. Mhango et al., demonstrated that while legume adoption and residue incorporation to support maize cropping was limited in sub-Saharan Africa, on-farm, participatory experimentation and research – along with nutritional education, introduction of new legume species and provision of extension services through the Soil, Food and Healthy

Communities project (http://soilandfood.org) – increased legume adoption in the participating Malawian communities. Thus, significant new knowledge was created when researchers where sensitive to psychological nuances and strategies that engender 'buy in' from farmers.

Food Security

PR related to food security has predominantly focused on the role of legumes in supporting the production of staple food crops. For example, Kamara et al., (2008) used a participatory research and extension approach (PREA) to identify effective means for controlling the parasitic weed Striga hermonthica in maize crops. The participatory process involved 30 local communities and 228 farmers representing 193 farmer groups in identifying their own problems and testing potential solutions to them in their own fields, supported by researchers and extension workers. Results showed not only effective Striga control but maize productivity increases of over 200%. The involvement of local farmers and groups in the evaluation process, firstly, helped to confirm that Striga control can best be achieved using soyabean followed by Striga-resistant maize together with productivity-increasing management practices and, secondly, promoted farmer-to-farmer extension. Vital new knowledge was co-created at least in part because the research team facilitated the participants' awareness of themselves *as* a knowledge system *within* a knowledge system (Leeuwis & Pyburn, 2002). A follow-up participatory adoption assessment exercise indicated widespread adoption of new varieties and management practices, despite the need for increased labour, one of the key constraints to adoption identified in other studies (e.g., Kitch et al., 1998).

In a related study, Ngwira, Thierfelder, and Lambert (2012) aimed to improve food security of smallholder farming families through enhanced productivity of maize and grain legumes. PAR methods with the involvement of researchers and extension workers helped Ngwira et al., facilitate co-learning and co-innovation among farmers in Malawi. Using Conservation Agriculture practices, the result was the development of maize/legume intercropping systems that did not reduce the yield of the main crop (maize) but also resulted in bonus yield from the intercropped legume (including cowpea, pigeon pea and soybean). Again, the power of involving farmers and those who work closely with them was self-evident; such partnerships can be built up as support networks that remain strong even when the researchers have departed.

Table 4. An example of participatory research to determine the role of legumes in food security

Authors and study area	Bezner Kerr et al. (2007), Malawi
Legume study theme	Food security
Study Aim	Using participatory research to encourage legume diversification for small holders for food and soil fertility.
Participatory process	Farmer research teams (FRT) selected by villagers to carry out on farm research (with the support of researchers) for the broader community and surveys conducted as to the effectiveness of this method.
Aspects of participatory study success	Through participatory research they identified knowledge gaps and provided nutritional and residue management education to highlight the multifunctional benefits of different legume varieties.
Constraints to participatory study success	3000 farmers were involved in the study over 4 years, however, local capacity was not developed to continue to support behavioural changes beyond the life of the project.
Lessons to learn	Local capacity development needs to be included in participatory research activities to ensure positive behavioural change beyond the participant group and the lifetime of the project (Criteria 5).

A review and in-depth analysis of a Malawian case study (Snapp & Silim, 2002) illustrated that farmer experimentation and adoption of legumes to contribute to child nutrition can be fostered among even the most resource-poor smallholders (Bezner Kerr et al., 2007). Farmers, particularly women, chose edible legume intercrops such as pigeon pea and groundnut over the mucuna green manure systems. Households participating in the studies reported feeding significantly more edible legumes to their children compared with control households. The research process, which incorporated nutritional education, fostered discussions within households and communities, formed the basis for sustained adoption of legume based systems. Again, social structures are inevitably part of the solution and therefore should be involved in PAR.

Conclusion

The benefits to be derived from participatory research have been emphasised with support from the literature and theoretical perspectives which warrant further attention in agricultural science. In summary, it is advocated that research which aims to advance legume productivity will bring together agricultural, industrial and social partners representing each step of the value chain (input suppliers, farmers, processors, retailers, consumers) with physical and social scientists. Throughout the life of your project it is advised that you have clear mechanisms to allow ongoing cross-fertilisation of ideas, and immediate follow-up in co-planning and co-executing new work. This tailored multi-actor approach will ensure that outputs are targeted to the needs of end-users, and that the practical knowledge of partners is taken into account. Innovation generated with an interactive approach tends to deliver solutions that are well adapted to circumstances and easier to implement, since the participatory process speeds up acceptance and dissemination. Impact will stem from participants' behaviour change and maintenance, which has as a pre-condition heightened self-awareness (achieved via actor engagement/involvement in the process) and an increase in available coping strategies (co-created by experts and participants). Sustained and long term implementation of results will be achieved through the ownership of the research process by all those impacted by the resultant science. End-users and practitioners are involved, not as targets of study, but as contributors of the entrepreneurial skills and practical, local knowledge required for true innovation.

REFERENCES

Abraham, B., Araya, H., Berhe, T., Edwards, S., Gujja, B., Khadka, R. B., Koma, Y. S., Sen, D., Sharif, A., Styger, E., Uphoff, N. and Verma, A. (2014). The system of crop intensification: Reports from the field on improving agricultural production, food security, and resilience to climate change for multiple crops. *Agriculture and Food Security, 3, 4.*

Asfaw, S., Shiferaw, B., Simtowe, F. and Lipper, L. (2012). Impact of modern agricultural technologies on smallholder welfare: Evidence from Tanzania and Ethiopia. *Food Policy*, 37, 283–295.

Ashby, J. A. and Lilja, N. (2004). Participatory Research: Does it Work? Evidence from Participatory Plant Breeding. *New directions for a diverse planet: Proceedings of the 4th International Crop Science Congress*, 26th September – 1st october 2004. Brisbane, Australia.

Bandura, A. (1986). *Social foundations of thought and action*. Englewood Cliffs, NJ: Prentice-Hall.

Bandura, A. (1989). Human Agency in Social Congitive Theory. *American Psychologist*, 44, 1175-1184.

Barlett, P. F. (1980). *Agricultural Decision Making: Anthropological Contributions to Rural Development*. New York City, NY: Academic Press.

Bellows, A. C. and Hamm, M. W. (2001). Local autonomy and sustainable development: Testing import substitution in more localized food systems. *Agriculture and Human Values*, 18, 271-284.

Bezner Kerr, R., Snapp, S., Chirwa, M., Shumba, L. and Msachi, R. (2007). Participatory research on legume diversification with Malawian smallholder farmers for improved human nutrition and soil fertility. *Experimental Agriculture*, 43, 437-453.

Carberry, P., Gladwin, C. and Twomlow, S. (2004). Linking simulation modelling to participatory research in smallholder farming systems. In

R. J. Delve and M. E. Probert (Eds.), *Modelling Nutrient Management in Tropical Cropping Systems*. ACIAR Proceedings No. 114, Canberra, Australia, pp. 32–46.

Charmaz, K. (2006). *Constructing Grounded Theory: A Practical Guide Through Qualitative Analysis*. London, England: Sage.

Chevalier, J. M. and Buckles, D. (2013). *Participatory Action Research : Theory & Methods for Engaged Inquiry*. Oxon, England : Routledge.

Courty, P. E., Smith, P., Koegel, S., Redecker, D. and Wipf, D. (2014). Inorganic nitrogen uptake and transport in beneficial plant root-microbe interactions. *Critical Reviews in Plant Sciences*, 34, 4-16.

Darnhofer, I., Schneeberger, W. and Freyer, B. (2005). Converting or not converting to organic farming in Austria: Farmer types and their rationale. *Agriculture and Human Values*, 22, 39-52.

Defoer, T., De Groote, H., Hilhorst, T., Kante, S. and Budelman, A. (1998). Participatory action research and quantitative analysis for nutrient management in southern Mali: a fruitful marriage? *Agriculture, Ecosystems and Environment*, 71, 215–228.

De Bruijn, F. J. (2014). Biological Nitrogen Fixation. In B. Lugtenberg (Ed.), *Principles of Plant-Microbe Interactions: Microbes for Sustainable Agriculture*. Basel, Switzerland: Springer.

Durant, M. and Scarafoni, A. (2015). An open window onto the food of the near future: Can legume seed proteins be a valid alternative to animal sources to fulfil the growing need of food proteins? *Agro FOOD Industry Hi Tech*, 26, 8-10.

Edwards-Jones, G. (2006). Modelling farmer decision-making: Concepts, progress and challenges. *Animal Science*, 82, 783-790.

Elden, M., Reason, P. and Rowan, J. (1981). *Human inquiry: A sourcebook of new paradigm research*. Chichester, England: Wiley.

European Innovation Partnership (2014). *EIP-Agri Focus Group Protein Crops: Final Report*. Brussels, Belgium: European Commission.

Ghassali, F., Cocks, P. S., Osman, A. E. and Leybourne, M. (1999). Rehabilitation of degraded grasslands in north Syria: Use of farmer participatory research to encourage the sowing of annual pasture legumes. *Experimental Agriculture*, 35, 489-506.

Grundy, S. (1987). *Curriculum: Product or Praxis*. London, England: The Falmer Press.

Guba, E. G. (Ed.). *The paradigm dialog*. Newbury Park, CA: Sage.

Guerin, L. J. and Guerin, T. F. (1994). Constraints to the adoption of innovations in agricultural research and environmental management : A review. *Australian Journal of Experimental Agriculture*, 34(4), 549-571.

Home, R. and Rump, N. (2015). Evaluation of a multi-case participatory action research project: The case of SOLINSA. *The Journal of Agricultural Education and Extension*, 21, 73-89.

Horton, D. and Mackay, R. (2003). Using evaluation to enhance institutional learning and change: Recent experiences with agricultural research and development. *Agricultural Systems*, 78, 127-142.

Ingram, J., Mills, J., Dibari, C., Ferrise, R., Ghaley, B. B., Hansen, J. G., Iglesias, A., Karaczun, Z., McVittie, A., Merante, P., Molnar, A. and Sánchez, B. (2016). Communicating soil carbon science to farmers : Incorporating credibility, salience and legitimacy. *Journal of Rural Studies*, 48, 115-128.

Johnson, N., Lilja, N. and Ashby, J. A. (2003). Measuring the impact of user participation in agricultural and natural resource management research. *Agricultural Systems*, 78, 287-306.

Kamara, A. Y., Ellis-Jones, J., Amaza, P., Omoigui, L. O., Helsen, J., Dugje, I. Y., Kamai, N., Menkir, A. and White, R. W. (2008). A participatory approach to increasing productivity of maize through *Striga hermonthica* control in Northeast Nigeria. *Experimental Agriculture,* 44, 349–364.

Kirchhoff, C. J., Lemos, M. C. and Dessai, S. (2013). Actionable knowledge for environmental decision making: Broadening the usability of climate science. *Annual Review of Environment and Resources*, 38, 393-414.

Kitch, L. W., Boukar, O., Endondo, C. and Murdock, L. L. (1998). Farmer acceptability criteria in breeding cowpea. *Experimental Agriculture*, 34, 475-486.

Leeuwis, C. and Pryburn, R. (Eds.) (2002). *Wheelbarrow full of frogs: Social learning in rural resource management*. Assen, Netherlands: Koninklijke Van Gorcum.

Liu, E. M. (2013). Time to change what to sow: Risk preferences and technology adoption decisions of cotton farmers in China. *The Review of Economics and Statistics*, 95, 1386-1403.

Lyon, A., Bell, M. M., Gratton, C. and Jackson, R. (2011). Farming without a recipe: Wisconsin graziers and new directions for agricultural science. *Journal of Rural Studies*, 27, 384-393.

Meijer, S. S., Catacutan, D., Ajayi, O. C., Sileshi, G. W. and Nieuwenhuis, M. (2015). The role of knowledge, attitudes and perceptions in the uptake of agricultural and agroforestry innovations among smallholder farmers in sub-Saharan Africa. *International Journal of Agricultural Sustainability*, 13, 40-54.

Mekbib, F. (1997). Farmer participatory variety selection in beans. *Experimental Agriculture*, 33, 399-408.

Mhango, W. G., Snapp, S. S. and Phiri, G. Y. K. (2012). Opportunities and constraints to legume diversification for sustainable maize production on smallholder farms in Malawi. *Renewable Agriculture and Food Systems*, 28, 234-244.

Muro, M. and Jeffrey, P. (2008). A critical review of the theory and application of social learning in participatory natural resource management processes. *Journal of Environmental Planning and Management*, 51, 325-344.

Nederlof, E. S. and Dangbégnon, C. (2007). Lessons for farmer-oriented research: Experiences from a West African soil fertility management project. *Agriculture and Human Values*, 24, 369-387.

Neef, A. and Neubert, D. (2011). Stakeholder participation in agricultural research projects: A conceptual framework for reflection and decision-making. *Agriculture and Human Values*, 28, 179-194.

Ngwira, A. M. R., Thierfelder, C. and Lambert, D. M. (2012). Conservation agriculture systems for Malawian smallholder farmers: long-term effects on crop productivity, profitability and soil quality. *Renewable Agriculture and Food Systems*, 1-14.

Nyende, P. and Delve, R. J. (2004). Farmer participatory evaluation of legume cover crop and biomass transfer technologies for soil fertility improvement using farmer criteria, preference ranking and logit regression analysis. *Experimental Agriculture*, 40, 77-88.

Öhlmér, B., Olson, K. and Brehmer, B. (1998). Understanding farmers' decision making processes and improving managerial assistance. *Agricultural Economics*, 18, 273-290.

Pretty, J. N. (1995). Participatory Learning for Sustainable Agriculture. *World Development*, 23, 1247-1263.

Snapp, S. S., Blackie, M. J. and Donovan, C. (2003). Realigning research and extension to focus on farmers' constraints and opportunities. *Food Policy*, 28, 349-363.

Snapp, S. S. and Silim, S. N. (2002). Farmer preferences and legume intensification for low nutrient environments. *Plant & Soil*, 245, 181-192.

Song, Y. (1998). *'New'seed in 'old' China: impact of CIMMYT collaborative programme on maize breeding in south-western China.* Thesis (PhD). Wageningen, The Netherlands: Wageningen University.

Struik, P. C., Kuyper, T. W., Brussaard, L. and Leeuwis, C. (2014). Deconstructing and unpacking scientific controversies in intensification and sustainability: why the tensions in concepts and values? *Current Opinion in Environmental Sustainability*, 8, 80-88.

Tilman, D., Cassman, K. G., Matson, P. A., Naylor, R. and Polasky, S. (2002). Agricultural sustainability and intensive production practices. *Nature*, 418, 671-677.

Van de Fliert, E. and Braun, A. R. (2002). Conceptualizing integrative, farmer participatory research for sustainable agriculture: From opportunities to impact. *Agriculture and Human Values*, 19, 25-38.

Weed, M. (2009). Research quality considerations for grounded theory research in sport and exercise psychology. *Psychology of Sport & Exercise*, 10, 502-510.

Whyte, W. F. (1989). Advancing Scientific Knowledge through Participatory Action Research. *Sociological Forum*, 4, 367-85.

Willock, J., Deary, I., McGregor, M., Sutherland, A., Edwards-Jones, G., Morgan, O., Dent, B., Grieve, R., Gibson, G. and Austin, E. (1999). Farmers' attitudes, Objectives, Behaviours, and Personality Traits: The Edinburgh Study of Decision Making on Farms. *Journal of Vocational Behaviour,* 54, 5-36.

Witcombe, J. R. (1999). Do farmer participatory methods apply more to high potential areas than to marginal ones? *Outlook on Agriculture*, 28, 43-49.

In: Legumes for Global Food Security ISBN: 978-1-53612-265-7
Editors: Alfonso Clemente et al.

Chapter 6

A SEED TRAIT STUDIED BY GREGOR MENDEL IN *PISUM SATIVUM* L. (PEA): POTENTIAL PREVENTION OF TYPE 2 DIABETES

Katerina Petropoulou[1], Louise Salt[2], Fred Warren[2], Claire Domoney[3,*], Peter Wilde[2] and Gary Frost[1]

[1]Nutrition and Dietetic Research Group, Section of Investigative Medicine, Faculty of Medicine, Hammersmith Campus, Imperial College London, UK
[2]Institute of Food Research, Norwich Research Park, Colney, Norwich, UK
[3]Department of Metabolic Biology, John Innes Centre, Norwich Research Park, Norwich, UK

---

* Corresponding Author Email: claire.domoney@jic.ac.uk.

ABSTRACT

Pulses have been attracting much interest in recent years from academia, industry and the general public. The attraction has been driven largely by the health benefits that pulses offer compared with refined, processed cereal based foods, coupled with the clear low environmental impact of these crops [1, 2]. Although the consumption of pulses is only a small fraction of that of staple crops such as wheat, maize and rice, the popularity of pulse foods is increasing, particularly in developed countries [3]. In the UK, consumer trends are moving towards a more varied diet, bringing in alternative sources of protein, carbohydrate and micronutrients and a move away from conventional cereals. Although some of these trends are based upon little or no rigorous scientific evidence, there are some clear health benefits associated with consumption of certain pulses. In this article, we will describe one such benefit that is the focus of a collaborative research programme which aims to improve our knowledge from basic plant science and genetics, through processing and consumption of foods and the physiological basis of associated health benefits. The programme aims to study how the type of starch contained in naturally occurring variants of pea seeds can help prevent the onset of type 2 diabetes. The specific aim is to understand how, through manipulating starch biosynthesis, the amount of fermentable carbohydrates reaching the bacterial community in the human colon might be increased. Fermentation of this carbohydrate is thought to produce metabolites which improve our ability to control blood sugar levels.

1. THE GLOBAL HEALTH CHALLENGE - NEED FOR A PUBLIC HEALTH SOLUTION

The World Health Organization (WHO) states that the global prevalence of type 2 diabetes has risen from 4.7% in 1980 (108 million patients) to 8.5% in 2014 (422 million patients) [4]. Interestingly, diabetes prevalence is a growing problem not only in developed countries, but also in developing middle- and low-income countries. The rising prevalence of type 2 diabetes and the ensuing economic burden of this chronic health condition across the globe calls for an urgent public health solution:

maintaining glucose homeostasis throughout life should be a public health priority.

Obesity is a major driver of insulin resistance and type 2 diabetes. However, not all people who are overweight or obese go on to develop type 2 diabetes. Recent evidence suggests that type 2 diabetes develops and progresses against a background of worsening insulin resistance (reduced cellular and tissue insulin sensitivity), developing alongside abnormalities in beta (β) cell function [5, 6]. In healthy individuals, β cells of the pancreas secrete insulin in a pulsatile way, at a periodicity of 5-15 min, with the aim of maintaining blood glucose levels within the tight range associated with metabolic health [7, 8]. Juhl *et al.* [9] have shown that the overall contribution of insulin pulsatility (basal insulin) accounts for at least 75% of the total insulin released. Therefore, disruption of insulin pulsatility can possibly cause the development of type 2 diabetes. McGarry's work in the 1990s suggested that an increase in lipid within the β cell disrupts insulin secretion [10]. This has been reinforced by Schofield *et al.* [6] whose model brings together lipotoxicity on the β cell, abnormal insulin release, insulin resistance and the development of type 2 diabetes.

Identifying dietary tools that can prevent disordered insulin secretion, arrest β cell decline and have an impact on insulin sensitivity can work as a strategy to combat the onset of type 2 diabetes. Recent evidence suggests that dietary fibre and resistant starches, which are fermented in the colon by the resident bacteria, lead to potential positive changes in both insulin resistance and β cell function [11-16]. Legume crops and pulses offer a high amount of fermentable carbohydrates in their seeds and can be used as a dietary tool in the prevention of type 2 diabetes.

2. Legumes, Pulses and Glucose Homeostasis

Dietary pulses are the edible non-oilseeds of leguminous crops such as peas, beans, chickpeas and lentils, which belong to the Leguminosae family [17, 18]. They offer significant nutritional as well as health advantages mainly due to the fact that they are rich in dietary fibre, they contain a high-quality vegetable protein and they have a low glycemic index as well as low energy density [19]. The glycemic index (GI) of pulses typically ranges from ~29 to 48; these are some of the lowest GI crops compared to other carbohydrate-containing food products [19]. They also contain a significant amount of resistant starches that cannot be digested by amylolytic enzymes in the small intestine and consequently these starches become available for fermentation by gut microbiota.

Many world health organizations, such as the European and American Diabetes Association [20, 21], suggest that pulses are ideal foods for optimizing health and preventing the onset of both type 2 diabetes and cardiovascular disease (CVD). Evidence from randomized control trials has shown that consumption of dietary pulses improves glycemic control [22, 23]. Jenkins *et al.* [24] fed volunteers with a 50g portion of available carbohydrates from eight different types of dried pulses and twenty-four common foods such as cereal, pasta, and biscuits. Results showed that the consumption of dried pulses led to a 45% lower mean peak rise in blood glucose concentration and mean area under the glucose curve, suggesting potentially valuable effects of carbohydrates from dried leguminous seeds on glucose metabolism [24]. Results from a systematic review of people with and without diabetes, including eleven trials examining the effect of dietary pulses alone on glycemic control, showed that fasting blood glucose and insulin concentrations were reduced by -0.82 (95% confidence interval (CI) -1.36 to -0.27) and -0.49 (95% CI -0.93 to -0.04), respectively [23]. Nineteen trials examining the effects of pulse consumption, incorporated in low glycemic index diets, showed a reduction in glycated haemoglobin (HbA_{1c}) of -0.28 (95% CI -0.42 to -0.14) [23]. Dietary pulses included in high fibre diets (eleven trials in total) lowered fasting blood glucose by -0.32 (95% CI -0.49 to -0.15) and HbA_{1c} by -0.27 (95% CI - 0.45 to – 0.09) [23].

From a mechanistic point of view there are several factors which might explain why pulses affect post-prandial glycaemia and most of them are attributed to the food structure. These factors will be explained below.

3. STARCH DIGESTIBILITY

The most abundant carbohydrate in pulse seeds is starch and starches occur in plant cells as semi-crystalline structures surrounded by a protein/lipid matrix [25]. Starches consist of two structural forms: amylose and amylopectin [25]. Amylose is formed from long linear chains of α-D-(1-4)-linked glucose monomers. Amylopectin is also largely formed from linear α-D-(1-4)-linked chains, but there are also side chains formed from additional α-D-(1-6) bonds resulting in a highly branched structure [25]. The ratio of amylose to amylopectin can vary greatly between plant species. It is influenced by both genetic and environmental factors and is one of the properties of starch that influences the rate of digestion and thus glucose release. Pulses with starches that have a higher amylose to amylopectin ratio tend to be digested more slowly in the small intestine, leaving a proportion of non-hydrolyzed starch that arrives intact in the colon for fermentation by gut microbiota [26].

Digestibility is also related to the starch granule structure of different botanical species. Starch granules consist of concentric rings of amorphous and pseudo-crystalline structures developed during starch granule growth. Starch in crystalline regions is highly organized with closely-packed, short branches, whereas starch in the amorphous region contains branch points in the amylopectin molecule, and is also interspersed with amylose [27]. Crystalline regions tend to be more resistant to amylolytic enzymes than amorphous regions [27]. Additionally, protein and lipid in the granule surface area can act as a barrier against hydrolysis and enzymatic digestion.

Digestibility can also be affected by alterations during processing such as soaking time, cooking, and thermal history. Hydrothermal treatment makes starch granules gelatinize, a process whereby starch granules swell and lose their crystalline structure. Amorphous regions swell and cause the crystalline structures to separate [27]. This swelling and opening up of the starch structure increases the accessibility of the amylolytic enzymes to the glyosidic bonds, thus making the starch more digestible, and releasing hydrolysis products more quickly. Starch gelatinization and swelling can be influenced by different factors. The disrupted granule structure of high amylose mutant starches, such as that of the pea *r* mutant (see later), means that the granule is unable to go through a coordinated gelatinization at a defined temperature, leading to a proportion of the starch not fully gelatinizing until well above 100°C. This 'high melting temperature' starch is far more resistant to amylase activity [28, 29]. Water availability is an important criterion here. If there is insufficient water available to hydrate the gelatinized starch, then this will restrict the amount of gelatinization, and thus digestibility. For example, starch that is trapped within a matrix or an intact plant cell may not fully gelatinize upon heating. Legume seeds often have thick and resistant cell walls and, if these remain intact following heating, the densely packed starches in the cell do not have enough water or space to swell and gelatinize thus limiting digestibility [30, 31]. Finally, when cooked starches are cooled they will recrystallize. This is known as retrogradation. Starches with few branched chains such as amylose recrystallize more readily and become more resistant to amylolytic enzymes and digestion. This is why starchy foods such as potatoes and pasta that have been cooked and then cooled have a much lower glycaemic index than the cooked form [32, 33].

4. Mendel, Pea Seeds and Starch Synthesis

The plant genetics community has recently celebrated 150 years of Mendelian genetics, whereby the fundamental principles of genetics were established by the careful and systematic studies of a monk, Gregor Mendel, using pea plants grown and multiplied within the gardens of a monastery in Brno, Moravia. Mendel presented his work and subsequently published the associated paper in 1865 and 1866, respectively, anniversaries of which have been marked by conferences in Brno and by a number of recent publications [34-36]. In parallel, the UN-declared International Year of Pulses in 2016 led to world-wide initiatives to promote the use of pulse crops, including pea, as an essential component of strategies to provide healthy food to feed a future world population and to mitigate climate change in the context of a global sustainable agriculture.

Fundamentally, the molecular basis for the variation in some of the traits studied by Mendel in pea is now understood [37-39] and their associated phenotypes are acquiring new relevance in terms of human health promotion. Specifically, the mutation leading to a loss of pigment in the flower and seed coat of pea impacts on the production of anthocyanins, compounds increasingly associated with a range of health benefits in humans. The genetic change in seed shape leading from a round to a wrinkled seed phenotype in pea, known as the *r* genetic locus, has been shown to be the consequence of an insertion event in the gene encoding an isoform of starch-branching enzyme (*SbeI*). Starch is the most abundant carbohydrate in pulse seeds, which accounts for 22-45% of the dry matter [40]. Starches are stored in plant-based tissues as insoluble, semi-crystalline granules and they have two structural polysaccharides: amylose and amylopectin [41]. Amylose and amylopectin are stored within amyloplasts (specialized plastids) [25]. As discussed in the last section, amylose is a linear or slightly branched polymer in which glucose residues are α-D-(1-4)-linked, whereas amylopectin is a larger branched molecule with α-D-(1-4) and α-D-(1-6) linkages [42]. SbeI is the major enzyme responsible for the amylose: amylopectin ratio found in wild-type pea seeds and, in mutant *sbeI* wrinkled pea seeds, starch is less highly branched with a higher amylose: amylopectin ratio. This structural change has important consequences for the shape of the starch granule (Figure 1) and the behavior of starch, most notably in its relative resistance to digestion. As discussed above, there is much interest in resistant starch in foods and its impact on slowing the release of sugars into the bloodstream, a process that has positive consequences in dealing with a global health problem, Type 2 diabetes. Hence there is now an opportunity to link fundamental plant genetic studies in crop plants, initiated by Mendel in the 1800s with the challenge of combatting the causes and consequences of a major human disease, which affects nutritional status in many parts of the world as a variant form of malnutrition associated with obesity, often co-existing alongside undernourishment in populations.

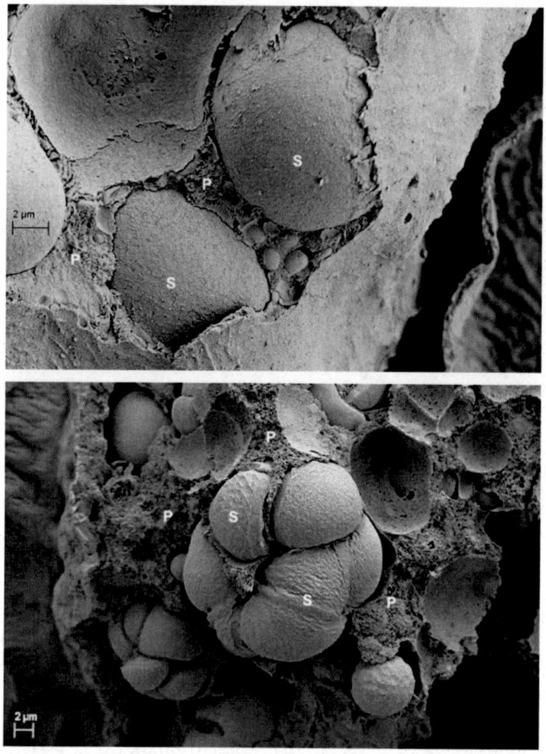

Figure 1: Scanning electron micrographs of fixed and fractured wild-type (RR, top panel) and mutant (rr, lower panel) pea seeds showing the structure of simple and compound starch grains, respectively. S, starch; P, protein; bars indicate 2μm scale.

5. Starch Biosynthesis and Genetics in Pea

There has been a long-term interest in starch biosynthetic mutations and their consequences for food quality. Many lesions in the pathway lead to higher accumulation of sucrose and derived metabolites in seeds and other plant parts, and have been exploited by the food industry in developing food products with higher consumer acceptability. The wrinkled-seeded phenotype of pea seeds, studied by Mendel, has therefore not only led to a food product with broad consumer appeal due to taste but also one which offers potential for delivery of greater amounts of resistant starch in foods. The purpose of our multi-disciplinary study outlined below is to determine to what extent the latter benefits may be realized in developing foods that help to combat a major disease. These studies have been aided greatly by the availability of near-isogenic lines that differ in the nature of their *SbeI/sbeI* allele and hence the concentration of resistant starch in their seeds.

Besides the gene insertion event that was revealed in the *sbeI* gene [37], other mutations may exist naturally in this gene and/or in other genes within the starch biosynthetic pathway. A second mutation has been reported to occur in another gene within pea lines originating from USA; this, the *rb* locus, affects the activity of the large subunit of ADP-glucose pyrophosphorylase [43]. A population of pea was mutagenized using ethyl methanesulfonate and shown to contain a range of mutations associated with the characteristic change in seed shape (wrinkling), which was used as a phenotypic screen of the population. Several distinct mutations affecting the proteins already described (SbeI and ADP-glucose pyrophosphorylase) were identified, as well as a range of mutant alleles affecting the activities of phosphoglucomutase, sucrose synthase and starch synthase at *rug 3*, *rug 4* and *rug 5* loci, respectively [44, 45].

The mutations at the *rug 3* locus were found to lead to the most severe reductions in the concentration of starch in the seed, lowering it to almost negligible amounts in the case of the most extreme mutant allele. These *rug 3* mutants had extremely high sugar concentrations, due to the lack of starch synthesis, and yielded 'super sweet' pea seeds [46]. The structure of the seed starch differs with all these discrete mutations [28] but, of these, the *r* and *rug 5* mutations might be predicted to impact positively on resistance to digestion as a positive attribute to combat type 2 diabetes.

6. Pea Starch Variation and Digestibility *In Vitro* and *In Vivo*

Starches are situated in both the seed coat and the cotyledon cells of pea [47]. The seed coat, which is the outer fibre, is commonly known as the hull and contains largely water-insoluble polysaccharides such as cellulose [47]. The cotyledon cell walls, known as the inner fibre, are composed of cellulose, hemicelluloses and pectin [48-50]. Round and wrinkled pea seeds show very different starch granular morphologies, in addition to their morphological differences reflected in both the wrinkled cotyledons and seed coat [51]. The natural variation for starch biosynthesis in pea seeds [52] discussed above is reflected in their overall starch composition [48]. Ratnayake at al. [51] reported that the amylose concentration in the starch granules from wild-type pea seeds can vary from 30-40%, whereas that in the starch granules from wrinkled pea seeds can vary from 60-70%.

The digestibility of starches depends mainly on the amylose to amylopectin ratio and starch granule structure as discussed above, while also being influenced by a number of factors, such as plant species, additional genetic factors and agronomy. Digestibility is then further influenced by processing conditions, physicochemical characteristics, and plant tissue microstructure and composition [53]. These factors all need careful consideration and detailed study in order to clarify and substantiate the effects of legume crops on lowering postprandial glucose levels, attributed mainly to their amount of resistant starches [54, 55].

Combining both *in vitro* and *in vivo* studies gives a deeper mechanistic understanding of the health benefits associated with consumption of plant foods with high resistant starch. Dietary intervention studies have shown the impact of foods rich in resistant starch on blood sugar and insulin sensitivity [25]. Short term *in vivo* studies examining the effect of pea consumption on glucose have shown a decrease in post-prandial glycaemia [55, 56]. Whether sustained consumption of pea seeds or their products has a direct effect on chronic glycemic conditions is not known. Nonetheless, Feskens *et al.* [57], in a 30-year follow-up study on incidence of type 2 diabetes, suggested that legume consumption was one of the factors linked with a reduction in the incidence of type 2 diabetes.

Complementary simulated digestion experiments can determine how the structural elements of the pea seed are affected by gastric and duodenal digestion and determine which products are produced following colonic fermentation. Botham *et al.* [26] showed that retrograded starch increased the amount of indigestible starch in ileostomy patients. Skrabanja *et al.* [58] compared the *in vitro* digestibility of different pea genotypes and showed that the high amylose pea genotypes contained the most resistant starch and had the slowest rate of digestion.

The structural features of pea starch granules have been studied and shown to have interesting features. Studies using atomic force microscopy revealed fine structural differences in the distribution of amylose and amylopectin within the starch granules of wild-type and mutant pea seeds, where the high amylose content can sometimes completely disrupt the overall structure of the starch granule [59, 60]. What impact these structural differences might have on the processing and digestibility of the starches remains to be established.

As discussed above, starch digestibility is related to the structure of starch granules, which in turn influences how the starches behave on heating in the presence of water [28]. During food processing, such as boiling, starches swell and as a result their molecular organization breaks down, crystallinity is lost and the disrupted granule structure becomes readily accessible to digestive enzymes [61]. Gelatinization usually occurs at temperatures between 55 and 70°C [62], leading to an open and disordered structure, associated with access to glycosidic bonds and the amylolytic activity of digestive enzymes. Studies have shown that starch types can behave differently to the thermodynamic parameters during gelatinization [28].

Retrogradation (see earlier) is also influenced by starch structure. The rate of crystallization upon cooling will be slower for amylopectin-rich starches but this will also depend on the structure of the amylopectin, for example the degree of branching and chain length [63]. Therefore, starches undergoing retrogradation, and more especially starches that contain a higher amount of amylose, are more resistant to digestive enzymes in the small intestine.

7. Health Outcomes: Resistant Starch and Colonic Health

In general, pea seeds are considered to act as a rich source of protein and dietary fibre (approximately 25% and 20% of dry weight, respectively) [64]. As with other fruits and vegetables, dietary fibre improves gastrointestinal health; for example, results from randomized controlled trials have shown that consumption of dietary fibre reduces the risk of CVD and colorectal cancer, decreases transit time and increases fecal bulk [65-67]. Pea seeds also contain a range of vitamins, minerals and phytochemicals, such as folates phenolics, phytates, saponins and oxalates [47], which are linked to a range of health benefits. The high dietary fibre and resistant starch content of pulses have been shown to significantly reduce post-prandial glucose response in comparison to other sources of starchy foods and to be a major contributor to the low glycemic index of pulses [68], raising the intriguing possibility that increased consumption of pulses could act as a successful dietary intervention to reduce the risk of type 2 diabetes. Moreover, increased consumption of pea and other legumes helps to directly address this issue by controlling the digestibility of starch and other macronutrients, either through their intrinsic resistance to digestion or indirectly by restricting accessibility of digestive enzymes to the starch embedded within the cellular matrix [69].

As resistant starches escape hydrolysis in the small intestine and pass *via* the digestive tract to the large intestine they are fermented and produce short chain fatty acids (SCFA). The principal SCFAs produced are acetate, propionate and butyrate and they are present in the approximate molar ratio of 60:20:20 [70]. The production of SCFAs through the fermentation process has been linked to a number of physiological effects that are beneficial to the host including: increased secretion of glucagon-like peptide 1 (GLP-1) (incretin effect), suppressed release of long chain free fatty acids, and direct effects of SCFAs on β cell function.

Studies in healthy individuals as well as patients with type 2 diabetes, who were following diets high in resistant starches, have shown improvements in post-prandial blood glucose as well as insulin sensitivity [11-16, 22, 71]. Long-term consumption of resistant starches has shown a positive effect on the concentrations of the insulin connecting peptide (C-peptide), peripheral insulin sensitivity and first phase insulin secretion [15, 22, 72]. These beneficial effects have been attributed to resistant starches, as constituents of the plant cell. Resistant starches reach the colon and are fermented by a myriad of gut bacteria, producing SCFAs and other molecules. SCFAs are natural ligands for the free fatty acid receptor 2 (FFAR2) and free fatty acid receptor 3 (FFAR3) [73]. These receptors are expressed in many cell types, including leukocytes, gut L-cells and pancreatic β cells [73]. Recent studies have shown that SCFAs induce the release of GLP-1 [74]. Specifically, experiments in primary colonic cell cultures have shown that SCFAs activate the secretion of GLP-1 through stimulation of FFAR2 [74, 75]. GLP-1 is an incretin gut hormone that stimulates insulin release after glucose ingestion and has been targeted as an anti-diabetic drug [76]. GLP-1 accounts for up to 60% of post-prandial insulin release in health individuals [77] and as a result this helps with lowering post-prandial glucose levels. In addition to this insulinotropic action, GLP-1 has been shown to prevent β cell decline by stimulating growth and proliferation, as well as by inhibiting apoptosis [78, 79].

Moreover, studies in humans have shown that resistant starch consumption reduces the release of circulating free fatty acids (FFA) from adipocytes and inhibits lipolysis [80, 81]. FFAs are known to play a significant role in glucose metabolism by causing insulin resistance and β cell decline. Long-term exposure to plasma FFA concentrations that stimulate insulin secretion could be "lipotoxic" to β cells [82]. However, further experiments *in vitro* have shown that SCFAs reduce lipolysis and the effect is mainly caused *via* stimulation of SCFA's natural ligands (FFAR2 and FFAR3) by acetate [80, 83-85]

Another possible mechanism relating SCFAs with β cell function is the expression of FFAR2 in human pancreatic islets. Pingitore *et al.* [86] showed that islet β cells express FFAR2 and this is possibly the reason for the effects of SCFAs on islet function [86]. They demonstrated that the SCFA, propionate, has beneficial effects on β cell function *in vivo*. The group has shown that, *in vitro*, propionate potentiates glucose-stimulated insulin release and maintains β cell mass *via* inhibition of apoptosis [86]. Therefore, these data suggest that SCFAs could play a significant role in β cell function through direct activation of the β cell FFAR2 receptor to improve glucose-stimulated insulin release. Current research is investigating the effect of pea seed starches on β cell pulsatility but the outcome is not yet known. However, it is a strong hypothesis that the high fermentable carbohydrate content of pea and legume seeds could improve glucose homeostasis through a number of direct and indirect physiological mechanisms.

8. Utilizing Pea Seeds in Everyday, Healthier Foods

Pea seeds can clearly form a very important part of a healthy diet, and the findings described in this chapter clearly make a strong case for increasing the proportion of pea (and other legumes) consumed within the human diet. With an increased understanding of the role of the gut microbiome in human health, not just in maintaining gut health, but also influencing broader health status, for example glycemic control and the development of type 2 diabetes, it is imperative to identify starchy foods which have slow digestion properties and are fermented to beneficial SCFA [87]. Combining such properties with the nutritional value conferred by other legume crop seed constituents presents a strong opportunity for developing health claims for pulse foods. The seeds of *r* mutant peas are an excellent candidate food ingredient to meet these criteria but, in order to optimize the slow digestibility and high fermentability of the mutant seeds and their products, it will be necessary to understand the mechanisms underlying the slow digestion of mutant starch. While pea is traditionally consumed in the UK in the form of whole, boiled, immature or mature seeds, its consumption is relatively limited and, to expand consumption, an obvious strategy would be to incorporate the seeds into foods in different forms, for example by using pea flour. Maintaining the beneficial functional properties of pea seeds in different food products will require an understanding of the key factors controlling their digestibility.

The digestion of starch from pea seeds is influenced by structures at a range of different scales, from the molecular scale structuring of individual starch chains, up to the structure of the whole intact pea seed, including structural features within individual starch granules (Figure 1), the behavior of starch within individual pea cells, and interactions with other non-starch components of the pea. Beneficial differences in melting temperature during gelatinization, and reduced granule swelling, in mutant relative to wild-type pea starch [29], will need to be preserved whether seeds are used in the form of a flour, or as purified starches. The matrix of plant tissues can be very important determinants of starch behavior, for example by limiting water availability, particularly in view of competition for water between starch and other cell components (e.g., protein) that may undergo thermal transitions at similar temperatures to starch, limiting gelatinization. The cell walls of the plant tissue can further limit starch gelatinization by reducing swelling of the starch granule, leading to starch which is significantly less digestible when cooked within a plant tissue [69].

For the purposes of the authors' ongoing collaborative project[†], we attempt to understand the effect of genetic variation in pea starch microstructure on preparation and cooking methods (milling or cooking) and to assess how processing in the presence or absence of the cellular matrix affects starch digestibility. Using a range of different processing techniques, combined with microstructural analysis and *in vitro* digestion models, we intend to identify the key structural features that are rate limiting in starch digestion in both *r* and wild-type pea seeds. The combined *in vitro* and *in vivo* analyses will enable the rational design of a range of food ingredients from both *r* and wild-type pea seeds, with starch digestion behavior which is optimized to provide the maximum health benefits in terms of slow starch digestion, and which is capable of providing high amounts of fermentable carbohydrates to the colon for fermentation to beneficial SCFA (Figure 2).

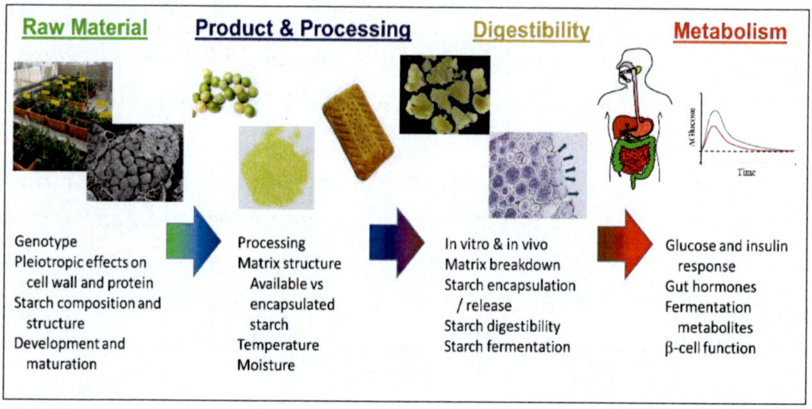

Figure 2: Schematic showing the authors' research programme activities from studies of the seeds from the chosen pea lines through to their use in human metabolic studies.

Conclusion

Current research is highlighting the role of pulses, including pea, in playing an important role in human glucose metabolism. The unique cell structure of seeds from legume crops, together with intrinsic properties of variant starches, can trigger several systems that enhance insulin sensitivity and beta cell function. Our understanding of the consequences of genetic defects in starch assembly in pea seeds provides opportunities to further enhance the beneficial effects of starches on human glucose metabolism. Deep mechanistic understanding of the relationship between plant cell structure and dietary glycemic control will help to provide evidence-based data to support health claims and to target crops that will reduce the future risk of development of type 2 diabetes. One of the major challenges is to increase the consumption of legumes in the UK population and globally. This challenge requires innovative food science, as well as a global strategy to promote pulse crops at all levels from farm to fork. The publication of a ten-year strategy for pulse crops, approved by the UN FAO, in December 2016 represents a major positive step in this direction [88].

ACKNOWLEDGMENTS

†The authors acknowledge Biotechnology and Biological Sciences Research Council (BBSRC) Diet and Health Research Industry Club (DRINC) for funding the project 'CRESTAR' (Crop Resistant Starches, grant reference BB/L025582/1). GF and KP are grateful for support from the MRC, BBSRC, an Integrative Mammalian Biology (IMB) Capacity Building Award, an FP7- HEALTH- 2009- 241592 EuroCHIP grant and the NIHR Biomedical Research Centre Funding Scheme. LS, FW and PW are grateful to the BBSRC for support to the Institute of Food Research through the Institute Strategic Programme Grant, Food and Health (BB/J004545/1). CD acknowledges support to the John Innes Centre from the BBSRC, through the Institute Strategic Programme MET (BB/J004561/1), the John Innes Foundation, the BBSRC and the Department for Environment, Food and Rural Affairs (Defra) for the LINK project (BBSRC: BB/H009787/1; Defra: LK09126; AHDB Horticulture: FV 351), and from Defra for funding the Pulse Crop Genetic Improvement Network (IF0147, CH0103).

REFERENCES

[1] Ha, V., Sievenpiper, J. L., de Souza, R. J., Jayalath, V. H., Mirrahimi, A., Agarwal, A., Chiavaroli, L., Mejia, S. B., Sacks, F. M., and Di Buono, M. (2014). Effect of dietary pulse intake on established therapeutic lipid targets for cardiovascular risk reduction: a systematic review and meta-analysis of randomized controlled trials. *Canadian Medical Association Journal 186*: E252-E262.

[2] Rebello, C., Greenway, F., and Finley, J. W. (2014). A review of the nutritional value of legumes and their effects on obesity and its related co-morbidities. *Obesity Reviews 15*: 392-407

[3] Bouchenak, M. and Lamri-Senhadji, M. (2013). Nutritional quality of legumes, and their role in cardiometabolic risk prevention: a review. *Journal of Medicinal Food 16*: 185-198.

[4] Organization, W. H., and Organization, W. H. (2011). Diabetes fact sheet No. 312. URL: http://www.who.int/mediacentre/factsheets /fs312/en/index.html.

[5] Ferrannini, E., Camastra, S., Gastaldelli, A., Sironi, A. M., Natali, A., Muscelli, E., Mingrone, G., and Mari, A. (2004). Beta-cell function in obesity effects of weight loss. *Diabetes 53*: S26-S33.

[6] Schofield, C., and Sutherland, C. (2012). Disordered insulin secretion in the development of insulin resistance and Type 2 diabetes. *Diabetic Medicine 29*: 972-979.

[7] Song, S. H., McIntyre, S. S., Shah, H., Veldhuis, J. D., Hayes, P. C., and Butler, P. C. (2000). Direct measurement of pulsatile insulin secretion from the portal vein in human Subjects 1. *The Journal of Clinical Endocrinology and Metabolism 85*: 4491-4499.

[8] Pørksen, N., Hollingdal, M., Juhl, C., Butler, P., Veldhuis, J. D., and Schmitz, O. (2002). Pulsatile insulin secretion: detection, regulation, and role in diabetes. *Diabetes 51*: S245-S254.

[9] Juhl, C., Grøfte, T., Butler, P. C., Veldhuis, J. D., Schmitz, O., and Pørksen, N. (2002). Effects of fasting on physiologically pulsatile insulin release in healthy humans. *Diabetes 51*: S255-S257.

[10] McGarry, J. D. (1994). Disordered metabolism in diabetes: have we underemphasized the fat component? J*ournal of Cellular Biochemistry 55*: 29-38.

[11] Robertson, M. D., Bickerton, A. S., Dennis, A. L., Vidal, H., and Frayn, K. N. (2005). Insulin-sensitizing effects of dietary resistant starch and effects on skeletal muscle and adipose tissue metabolism. *The American Journal of Clinical Nutrition 82*: 559-567.

[12] Nilsson, A., Östman, E., Preston, T., and Björck, I. (2008). Effects of GI vs content of cereal fibre of the evening meal on glucose tolerance at a subsequent standardized breakfast. *European Journal of Clinical Nutrition 62*: 712-720.

[13] Klosterbuer, A. S., Thomas, W., and Slavin, J. L. (2012). Resistant starch and pullulan reduce postprandial glucose, insulin, and GLP-1, but have no effect on satiety in healthy humans. *Journal of Agricultural and Food Chemistry 60*: 11928-11934.

[14] Kendall, C. W., Esfahani, A., Sanders, L. M., Potter, S. M., and Vidgen, E. (2010). The effect of a pre-load meal containing resistant starch on spontaneous food intake and glucose and insulin responses. *Journal of Food Technology 8*: 67-73.

[15] Bodinham, C., Smith, L., Thomas, E. L., Bell, J. D., Swann, J. R., Costabile, A., Russell-Jones, D., Umpleby, A. M., and Robertson, M. D. (2014). Efficacy of increased resistant starch consumption in human type 2 diabetes. *Endocrine Connections 3*: 75-84.

[16] Bodinham, C. L., Smith, L., Wright, J., Frost, G. S., and Robertson, M. D. (2012). Dietary fibre improves first-phase insulin secretion in overweight individuals. *PLoS One* 7: e40834.

[17] Hoover, R. and Sosulski, F. (1991). Composition, structure, functionality, and chemical modification of legume starches: a review. *Canadian Journal of Physiology and Pharmacology 69*: 79-92.

[18] Li, S. S., Kendall, C. W., Souza, R. J., Jayalath, V. H., Cozma, A. I., Ha, V., Mirrahimi, A., Chiavaroli, L., Augustin, L. S., and Blanco Mejia, S. (2014). Dietary pulses, satiety and food intake: A systematic review and meta analysis of acute feeding trials. *Obesity 22*: 1773-1780.

[19] McCrory, M. A., Hamaker, B. R., Lovejoy, J. C., and Eichelsdoerfer, P. E. (2010). Pulse consumption, satiety, and weight management. *Advances in Nutrition: An International Review Journal 1*: 17-30.

[20] Association, A. D. (2008). Nutrition recommendations and interventions for diabetes. *Diabetes Care 31*: S61-S78.

[21] Mann, J., De Leeuw, I., Hermansen, K., Karamanos, B., Karlström, B., Katsilambros, N., Riccardi, G., Rivellese, A., Rizkalla, S., and Slama, G. (2004). Diabetes and Nutrition Study Group (DNSG) of the European Association. Evidence based nutritional approaches to

the treatment and prevention of diabetes mellitus. *Nutrition, Metabolism and Cardiovascular Disease 14*: 373-394.

[22] Jenkins, D. J., Kendall, C. W., Augustin, L. S., Mitchell, S., Sahye-Pudaruth, S., Mejia, S. B., Chiavaroli, L., Mirrahimi, A., Ireland, C., and Bashyam, B. (2012). Effect of legumes as part of a low glycemic index diet on glycemic control and cardiovascular risk factors in type 2 diabetes mellitus: a randomized controlled trial. *Archives of Internal Medicine 172*: 1653-1660.

[23] Sievenpiper, J., Kendall, C., Esfahani, A., Wong, J., Carleton, A., Jiang, H., Bazinet, R., Vidgen, E., and Jenkins, D. (2009). Effect of non-oil-seed pulses on glycaemic control: a systematic review and meta-analysis of randomised controlled experimental trials in people with and without diabetes. *Diabetologia 52*: 1479-1495.

[24] Jenkins, D., Wolever, T., Taylor, R. H., Barker, H. M., and Fielden, H. (1980). Exceptionally low blood glucose response to dried beans: comparison with other carbohydrate foods. *British Medical Journal 281*: 578-580.

[25] Lovegrove, A., Edwards, C. H., De Noni, I., Patel, H., El, S. N., Grassby, T., Zielke, C., Ulmius, M., Nilsson, L., Butterworth, P. J., Ellis, P. R., and Shewry, P. R. (2017). Role of polysaccharides in food, digestion and health. *Critical Reviews in Food Science and Nutrition 57*: 237-253.

[26] Botham, R., Morris, V., Noel, T., and Ring, S. (1996). A study on the in vivo digestibility of retrograded starch. *Carbohydrate Polymers 29*: 347-352.

[27] Donald, A. (2004). *Understanding starch structure and functionality. In: Starch in food: Structure, function and applications* (Ed. A.-C. Eliasson), Woodhead Publishing Ltd., Cambridge, England, pp. 156-184.

[28] Bogracheva, T. Y., Cairns, P., Noel, T., Hulleman, S., Wang, T., Morris, V., Ring, S., and Hedley, C. (1999). The effect of mutant genes at the r, rb, rug3, rug4, rug5 and lam loci on the granular structure and physico-chemical properties of pea seed starch. *Carbohydrate Polymers 39*: 303-314.

[29] Tahir, R., Ellis, P. R., and Butterworth, P. J. (2010). The relation of physical properties of native starch granules to the kinetics of amylolysis catalysed by porcine pancreatic a-amylase. *Carbohydrate Polymers 81*: 57-62.

[30] Tovar, J., Bjoerck, I. M., and Asp, N. G. (1990). Analytical and nutritional implications of limited enzymic availability of starch in cooked red kidney beans. *Journal of Agricultural and Food Chemistry 38*: 488-493.

[31] Tovar, J., Granfeldt, Y., and Bjoerck, I. M. (1992). Effect of processing on blood glucose and insulin responses to starch in legumes. *Journal of Agricultural and Food Chemistry 40*: 1846-1851.

[32] Mishra, S., Hardacre, A., and Monro, J., *Food structure and carbohydrate digestibility*. 2012: INTECH Open Access Publisher.

[33] Najjar, N., Adra, N., and Hwalla, N. (2004). Glycemic and insulinemic responses to hot vs cooled potato in males with varied insulin sensitivity. *Nutrition Research 24*: 993-1004.

[34] Smýkal, P., Varshney, R. K., Singh, V. K., Coyne, C. J., Domoney, C., Kejnovský, E., and Warkentin, T. (2016). From Mendel's discovery on pea to today's plant genetics and breeding. *Theoretical and Applied Genetics 129*: 2267-2280.

[35] van Dijk, P. J. and Ellis, T. N. (2016). The full breadth of Mendel's genetics. *Genetics 204*: 1327-1336.

[36] Vollmann, J. and Buerstmayr, H. (2016). From phenotype to genotype: celebrating 150 years of Mendelian genetics in plant breeding research. *Theoretical and Applied Genetics 129*: 2237-2239.

[37] Bhattacharyya, M. K., Smith, A. M., Ellis, T. N., Hedley, C., and Martin, C. (1990). The wrinkled-seed character of pea described by Mendel is caused by a transposon-like insertion in a gene encoding starch-branching enzyme. *Cell 60*: 115-122.

[38] Hellens, R. P., Moreau, C., Lin-Wang, K., Schwinn, K. E., Thomson, S. J., Fiers, M. W., Frew, T. J., Murray, S. R., Hofer, J. M., and

Jacobs, J. M. (2010). Identification of Mendel's white flower character. *PloS One 5*: e13230.

[39] Ellis, T. N., Hofer, J. M., Timmerman-Vaughan, G. M., Coyne, C. J., and Hellens, R. P. (2011). Mendel, 150 years on. *Trends in Plant Science 16*: 590-596.

[40] Hoover, R. and Ratnayake, W. (2002). Starch characteristics of black bean, chick pea, lentil, navy bean and pinto bean cultivars grown in Canada. *Food Chemistry 78*: 489-498.

[41] Wang, S., Sharp, P., and Copeland, L. (2011). Structural and functional properties of starches from field peas. *Food Chemistry 126*: 1546-1552.

[42] Sajilata, M. G., Singhal, R. S., and Kulkarni, P. R. (2006). Resistant starch–a review. *Comprehensive Reviews in Food Science and Food Safety 5*: 1-17.

[43] Hylton, C. and Smith, A. M. (1992). The rb mutation of peas causes structural and regulatory changes in ADP glucose pyrophosphorylase from developing embryos. *Plant Physiology 99*: 1626-1634.

[44] Wang, T. and Hedley, C. (1991). Seed development in peas: knowing your three 'r's'(or four, or five). Seed Science Research 1: 3-14.

[45] Casey, R., Domoney, C., Forster, C., Hedley, C., Hitchin, E., and Wang, T. (1998). The effect of modifying carbohydrate metabolism on seed protein gene expression in peas. *Journal of Plant Physiology 152*: 636-640.

[46] Harrison, C. J., Mould, R. M., Leech, M. J., Johnson, S. A., Turner, L., Schreck, S. L., Baird, K. M., Jack, P. L., Rawsthorne, S., and Hedley, C. L. (2000). The rug3 locus of pea encodes plastidial phosphoglucomutase. *Plant Physiology 122*: 1187-1192.

[47] Dahl, W. J., Foster, L. M., and Tyler, R. T. (2012). Review of the health benefits of peas (Pisum sativum L.). *British Journal of Nutrition 108*: S3-S10.

[48] Tosh, S. M. and Yada, S. (2010). Dietary fibres in pulse seeds and fractions: Characterization, functional attributes, and applications. *Food Research International 43*: 450-460.

[49] Guillon, F. and Champ, M.-J. (2002). Carbohydrate fractions of legumes: uses in human nutrition and potential for health. *British Journal of Nutrition 88*: 293-306.

[50] Reichert, R. D. and MacKenzie, S. L. (1982). Composition of peas (Pisum sativum) varying widely in protein content. *Journal of Agricultural and Food Chemistry 30*: 312-317.

[51] Ratnayake, W. S., Hoover, R., and Warkentin, T. (2002). Pea starch: composition, structure and properties—a review. *Starch-Stärke 54*: 217-234.

[52] Ridout, M. J., Parker, M. L., Hedley, C. L., Bogracheva, T. Y., and Morris, V. J. (2003). Atomic force microscopy of pea starch granules: granule architecture of wild-type parent, r and rb single mutants, and the rrb double mutant. *Carbohydrate Research 338*: 2135-2147.

[53] Ring, S. G., Gee, J. M., Whittam, M., Orford, P., and Johnson, I. T. (1988). Resistant starch: its chemical form in foodstuffs and effect on digestibility in vitro. *Food Chemistry 28*: 97-109.

[54] Cairns, P., Morris, V., Botham, R., and Ring, S. (1996). Physicochemical studies on resistant starch in vitro and in vivo. *Journal of Cereal Science 23*: 265-275.

[55] Hoover, R. and Zhou, Y. (2003). In vitro and in vivo hydrolysis of legume starches by a-amylase and resistant starch formation in legumes—a review. *Carbohydrate Polymers 54*: 401-417.

[56] Mollard, R., Zykus, A., Luhovyy, B., Nunez, M., Wong, C., and Anderson, G. (2012). The acute effects of a pulse-containing meal on glycaemic responses and measures of satiety and satiation within and at a later meal. *British Journal of Nutrition 108*: 509-517.

[57] Feskens, E. J., Virtanen, S. M., Räsänen, L., Tuomilehto, J., Stengård, J., Pekkanen, J., Nissinen, A., and Kromhout, D. (1995). Dietary factors determining diabetes and impaired glucose tolerance: a 20-year follow-up of the Finnish and Dutch cohorts of the Seven Countries Study. *Diabetes Care 18*: 1104-1112.

[58] Skrabanja, V., Liljeberg, H. G., Hedley, C. L., Kreft, I., and Björck, I. M. (1999). Influence of genotype and processing on the in vitro

rate of starch hydrolysis and resistant starch formation in peas (Pisum sativum L.). *Journal of Agricultural and Food Chemistry 47*: 2033-2039.

[59] Parker, M. L., Kirby, A. R., and Morris, V. J. (2008). In situ imaging of pea starch in seeds. *Food Biophysics 3*: 66-76.

[60] Ridout, M. J., Parker, M. L., Hedley, C. L., Bogracheva, T. Y., and Morris, V. J. (2006). Atomic force microscopy of pea starch: Granule architecture of the rug3-a, rug4-b, rug5-a and lam-c mutants. *Carbohydrate Polymers 65*: 64-74.

[61] Cummings, J. H., and Englyst, H. N. (1995). Gastrointestinal effects of food carbohydrate. *The American Journal of Clinical Nutrition 61*: 938S-945S.

[62] Jenkins, P. J. and Donald, A. M. (1998). Gelatinisation of starch: a combined SAXS/WAXS/DSC and SANS study. *Carbohydrate Research 308*: 133-147.

[63] Gudmundsson, M. (1994). Retrogradation of starch and the role of its components. *Thermochimica Acta 246*: 329-341.

[64] de Almeida Costa, G. E., da Silva Queiroz-Monici, K., Reis, S. M. P. M., and de Oliveira, A. C. (2006). Chemical composition, dietary fibre and resistant starch contents of raw and cooked pea, common bean, chickpea and lentil legumes. *Food Chemistry 94*: 327-330.

[65] Threapleton, D. E., Greenwood, D. C., Evans, C. E., Cleghorn, C. L., Nykjaer, C., Woodhead, C., Cade, J. E., Gale, C. P., and Burley, V. J. (2013). Dietary fibre intake and risk of cardiovascular disease: systematic review and meta-analysis. *The BMJ 347*: f6879.

[66] Butler, L., Wang, R., Koh, W., and Yu, M. (2008). Prospective study of dietary patterns and colorectal cancer among Singapore Chinese. *British Journal of Cancer 99*: 1511-1516.

[67] Cummings, J. H., Bingham, S. A., Heaton, K. W., and Eastwood, M. A. (1992). Fecal weight, colon cancer risk, and dietary intake of nonstarch polysaccharides (dietary fiber). *Gastroenterology 103*: 1783-1789.

[68] Seewi, G., Gnauck, G., Stute, R., and Chantelau, E. (1999). Effects on parameters of glucose homeostasis in healthy humans from

ingestion of leguminous versus maize starches. *European Journal of Nutrition 38*: 183-189.

[69] Edwards, C. H., Grundy, M. M., Grassby, T., Vasilopoulou, D., Frost, G. S., Butterworth, P. J., Berry, S. E., Sanderson, J., and Ellis, P. R. (2015). Manipulation of starch bioaccessibility in wheat endosperm to regulate starch digestion, postprandial glycemia, insulinemia, and gut hormone responses: a randomized controlled trial in healthy ileostomy participants. *The American Journal of Clinical Nutrition 102*: 791-800.

[70] Cummings, J., Pomare, E., Branch, W., Naylor, C., and Macfarlane, G. (1987). Short chain fatty acids in human large intestine, portal, hepatic and venous blood. *Gut 28*: 1221-1227.

[71] Klosterbuer, A. S., Hullar, M. A., Li, F., Traylor, E., Lampe, J. W., Thomas, W., and Slavin, J. L. (2013). Gastrointestinal effects of resistant starch, soluble maize fibre and pullulan in healthy adults. *British Journal of Nutrition 110*: 1068-1074.

[72] Robertson, M. D., Wright, J. W., Loizon, E., Debard, C., Vidal, H., Shojaee-Moradie, F., Russell-Jones, D., and Umpleby, A. M. (2012). Insulin-sensitizing effects on muscle and adipose tissue after dietary fiber intake in men and women with metabolic syndrome. *The Journal of Clinical Endocrinology and Metabolism 97*: 3326-3332.

[73] Nøhr, M. K., Pedersen, M. H., Gille, A., Egerod, K. L., Engelstoft, M. S., Husted, A. S., Sichlau, R. M., Grunddal, K. V., Seier Poulsen, S., and Han, S. (2013). GPR41/FFAR3 and GPR43/FFAR2 as cosensors for short-chain fatty acids in enteroendocrine cells vs FFAR3 in enteric neurons and FFAR2 in enteric leukocytes. *Endocrinology 154*: 3552-3564.

[74] Tolhurst, G., Heffron, H., Lam, Y. S., Parker, H. E., Habib, A. M., Diakogiannaki, E., Cameron, J., Grosse, J., Reimann, F., and Gribble, F. M. (2012). Short-chain fatty acids stimulate glucagon-like peptide-1 secretion via the G-protein–coupled receptor FFAR2. *Diabetes 61*: 364-371.

[75] Psichas, A., Sleeth, M., Murphy, K., Brooks, L., Bewick, G., Hanyaloglu, A., Ghatei, M., Bloom, S., and Frost, G. (2015). The

short chain fatty acid propionate stimulates GLP-1 and PYY secretion via free fatty acid receptor 2 in rodents. *International Journal of Obesity 39*: 424-429.

[76] Behme, M. T., Dupré, J., and McDonald, T. J. (2003). Glucagon-like peptide 1 improved glycemic control in type 1 diabetes. *BMC Endocrine Disorders 3*: 3.

[77] Salehi, M., Aulinger, B., Prigeon, R. L., and D'Alessio, D. A. (2010). Effect of endogenous GLP-1 on insulin secretion in type 2 diabetes. *Diabetes 59*: 1330-1337.

[78] Yabe, D. and Seino, Y. (2011). Two incretin hormones GLP-1 and GIP: comparison of their actions in insulin secretion and ß-cell preservation. *Progress in Biophysics and Molecular Biology 107*: 248-256.

[79] Vilsbøll, T. (2009). The effects of glucagon-like peptide-1 on the beta cell. *Diabetes, Obesity and Metabolism 11*: 11-18.

[80] Ge, H., Li, X., Weiszmann, J., Wang, P., Baribault, H., Chen, J. L., Tian, H., and Li, Y. (2008). Activation of G protein-coupled receptor 43 in adipocytes leads to inhibition of lipolysis and suppression of plasma free fatty acids. *Endocrinology 149*: 4519-4526.

[81] Higgins, J. A., Higbee, D. R., Donahoo, W. T., Brown, I. L., Bell, M. L., and Bessesen, D. H. (2004). Resistant starch consumption promotes lipid oxidation. *Nutrition and Metabolism 1*: 1-11.

[82] Kim, J. W. and Yoon, K. H. (2011). Glucolipotoxicity in pancreatic ß-cells. *Diabetes and Metabolism Journal 35*: 444-450.

[83] Asp, N. G., Björck, I., Holm, J., Nyman, M., and Siueström, O. (1987). Enzyme resistant starch fractions and dietary fibre. *Scandinavian Journal of Gastroenterology 22*: 29-32.

[84] Hong, Y. H., Nishimura, Y., Hishikawa, D., Tsuzuki, H., Miyahara, H., Gotoh, C., Choi, K. C., Feng, D. D., Chen, C., and Lee, H. G. (2005). Acetate and propionate short chain fatty acids stimulate adipogenesis via GPCR43. *Endocrinology 146*: 5092-5099.

[85] Wolever, T., Spadafora, P., and Eshuis, H. (1991). Interaction between colonic acetate and propionate in humans. *The American Journal of Clinical Nutrition 53*: 681-687.

[86] Pingitore, A., Chambers, E. S., Hill, T., Maldonado, I. R., Liu, B., Bewick, G., Morrison, D. J., Preston, T., Wallis, G. A., and Tedford, C. (2016). The diet-derived short chain fatty acid propionate improves beta-cell function in humans and stimulates insulin secretion from human islets in vitro. *Diabetes, Obesity and Metabolism.*

[87] Byrne, C., Chambers, E., Morrison, D., and Frost, G. (2015). The role of short chain fatty acids in appetite regulation and energy homeostasis. International Journal of Obesity 39: 1331-1338.

[88] Ten-year research strategy for pulse crops, International year of pulses. December 2016. URL: http://iyp2016.org/index.php?option =com_content&view=article&id=275.

In: Legumes for Global Food Security
Editors: Alfonso Clemente et al.
ISBN: 978-1-53612-265-7

Chapter 7

THE NUTRITIONAL AND NUTRACEUTICAL VALUES OF SOME "READY-TO-EAT" EXPANDED LEGUME PRODUCTS

Mercedes M. Pedrosa[*] and Carmen Cuadrado
Departamento de Tecnología de Alimentos,
SGIT-INIA, Madrid, Spain

ABSTRACT

Legumes occupy an important place in human nutrition, since they are a cheap and valuable source of protein and an important constituent of the diet in many countries. From a nutritional point of view, beans and peas are classified by the USDA in both the vegetable and protein food group, and these foods are excellent sources of plant proteins, as well as, slowly assimilated carbohydrates, minerals and fiber. Furthermore, they contain different bioactive compounds related to the prevention or reduction of the risk of chronic diseases, such as coronary heart disease, type II diabetes and obesity. Nowadays, pulses are gaining more interest for the development of healthy and functional foods. Extrusion cooking is

[*] Corresponding Author Email: mmartin@inia.es.

a high-temperature, short-time process used by the industry for the production of a wide variety of products like pasta, breakfast cereals, baby foods, snack foods, texturized proteins, etc. Consumer's acceptance of extruded foods is mainly due to the convenience, value, attractive appearance and texture found for these products. Extrusion of legume flours or legume-cereal blends promoted important changes in both nutritive and bioactive compounds present in the dry legumes and they accounted for the nutritional and nutraceutical quality of finished products. So, legumes are highly feasible for the development of high-nutritional, low-calorie "ready-to-eat" products.

Keywords: legumes, extrusion processing, bioactive compounds, nutritive value, functional foods

1. INTRODUCTION

Legumes and pulses represent one of the most important components of the human diet and have been consumed for at least 10000 years through the world. According to the FAO (2004) definition, the term "pulses" includes 11 primary pulses (beans, broad beans, peas, chickpeas, black-eyed peas, pigeon peas, lentils, bambara groundnut, vetch, lupins and other minor pulses) and excludes legumes used for oil extraction (soybean and peanut) and those harvested green for food (green pea, green bean). In Europe, higher consumption occurs in the Mediterranean area, and there is high variability in the species and varieties consumed by humans (DAFNE 2009). Pulses are recommended as part of a healthy diet by different government agencies, for example the USDA recommends the Americans consume between 2.5-3.5 cups of pulses per week (US Dietary 2010).

Legumes are rich in protein gluten-free, slowly assimilated carbohydrates, minerals (as calcium, iron and zinc), B and E vitamins, and fiber, and they are considered an ideal complement to cereals for vegetarians. However, they contain some compounds, generally known as anti-nutritional factors that impede the digestion and absorption of some of the nutritive components. Nowadays, these compounds are considered as nutritionally active factors or bioactive components (such as galactosides, phytates, saponins, protease inhibitors) that may play healthy roles; currently, they are receiving attention for being an excellent functional food and, in this context, consuming them could reduce the risk of chronic diseases, such as coronary heart disease, type II diabetes and obesity (Campos-Vega, Loarca-Piña, and Oomah 2010, Sievenpiper et al. 2009).

Although the nutritional and health benefits of legumes are well recognized globally and recommends as part of healthy eating, there is a downward trend in legume consumption all the world (FAOSTAT 2016). In the last decades, consumer's demands in the field of food production have changed considerably. Consumers believe that foods contribute directly to their health (Mollet and Rowland 2002) and foods are not intended to only satisfy hunger and to provide necessary nutrients but also to prevent diseases and improve well-being. Nowadays, there is an increase consciousness of health-associated value of pulses, however, their nutritional value is not always well understood and appreciated by consumers. This is due in part to the prolonged soaking and cooking processes required. Of the main factors that consumers consider when selecting food for cooking/eating at home (flavor, texture, nutrition, cost, safety, and convenience), the flavor has been shown to be the most important (Azarnia et al. 2011). Enhancing the flavor of legumes through preparation and cooking can increase the consumption of this healthy food, especially among children (Poelman, Delahunty, and de Graaf 2013) So, the development of "ready-to-eat" legume products, such as canned pulses and snacks, would favor the increase in the at-home intake and adapting it to current lifestyles because the traditional methods of preparation are generally labor-intensive and time-consuming.

There are different options for the inclusion of grain legumes in innovative food preparation with increased nutritional value and specific functional attributes. Examples include canned products or the fortification of foods such as pastries, breads, pasta, snacks, breakfast cereals, etc. (Boye, Zare, and Pletch 2010). These kind of products could contribute to the diversification of diets and increase legume consumption due to their higher convenience when compared to traditional preparations. Extrusion/cooking technology is a versatile and modern food processing that converts agricultural materials into fully cooked, shelf-stable food products with enhanced textural attributes and flavor (Alam et al. 2016). It is a continuous process with high productivity and significant nutrient retention, due to the high temperature and short time required in which moisture, expansive, starchy and/or proteinaceous food materials are plasticized and cooked by a combination of moisture, temperature and mechanical shear, resulting in molecular transformation and chemical reactions (Castells et al. 2005). Extrusion has found numerous applications, including different ready-to-eat cereals, snacks, and textured meat-like materials from defatted high protein flours or precooked food mixtures for infant formulas (Harper 1989). Usually, these products are elaborated with maize grits because of its high starch content, however, they are considered high energy density foods that promote weight gain and some diseases such as metabolic syndrome, cardiovascular events, hypertension, obesity, etc. Cereal-based snacks show a low nutritional value, mainly due to their poor protein quality. The incorporation of legumes represents an economic way to improve the biological value of proteins from cereal-based foods. Extrusion offers great potential for the development of many novel legume-fortified cereal products, although the use of combinations of cereals and legumes requires the control of the processing parameters to obtain products with optimum chemical, physical and functional characteristics. Extrusion processing has been reported to be an effective method to retain nutritional components and to improve protein digestibility in legumes. In addition, the extrusion process can reduce or inactivate some anti-nutritional factors present in legumes, such as trypsin inhibitors, lectins, phytates or phenols, and improve starch and protein digestibility (Alonso, Aguirre, and Marzo 2000, Alonso, Orúe, and Marzo 1998). For the production of nutritious acceptable snacks, rates of starch fortification seem to vary depending on the material used. Legumes have been reported to cause good expansion and are regarded as highly feasible for the development of high-nutritional, low-calorie snacks (Berrios 2006).

In the present chapter, we have summarized the changes on the nutritive and bioactive compounds of some "ready-to-eat" products based on extruded legumes.

2. Effect of Extrusion/Cooking on the Nutritive Value of Legumes

Extrusion cooking has been widely investigated to improve the nutritional quality of different vegetables, mainly cereals and recently legumes, vegetables and by-products. Extrusion of grain legumes can result in a "ready-to-eat" product with a better nutritional value compared to traditional snacks made from corn starch. The extrusion parameters (moisture, temperature, feed ratio, and residence time) must be carefully control in order to improve or maintain the quality of the foods, as well as to control the density, porosity and expansion rates that allowed to obtain an acceptable expanded ready-to-eat product. Increasing protein or fiber concentrations reduce the extrudate expansion ratio, water solubility and increased extrudate density (Allen, Carpenter, and Walsh 2007). Despite the increasing use of extrusion process, it still has to be mastered due its complexity, since the nature of materials and small variations in processing conditions affect response variables, as well as the quality of final products. Generally, mild extrusion conditions (<180°C, <15% moisture) favor higher protein and starch digestibilities, increased soluble dietary fiber, decreased lipid oxidation, higher retention of vitamins and higher absorption of minerals (Alonso et al. 2000, Korus, Gumul, and Achremowicz 2006, Singh, Gamlath, and Wakeling 2007, Stojceska et al. 2009)

In general, total protein changes very little during extrusion; aggregation and fragmentation reactions reduced the soluble protein content and an increase in protein digestibility is observed (Alonso, Orúe, and Marzo 1998, Ummadi, Chenoweth, and NG 1995, Zamora 2003). Varela et al. (2007) found a significant reduction of soluble protein content after extrusion of pea, kidney bean, faba bean, chickpea and lupin. This reduction was depended on the seed, so the higher reduction was observed in extruded faba bean seeds (97%), and the lowest one corresponded to extruded lupin seeds (65%). This decrease might be caused by the combined effect of intermolecular disulphide bond formation and hydrophobic interaction of legume proteins taking place during extrusion processing. Chen, Wei, and Zhang (2011) determined chemical-cross-linking and molecular aggregation of soybean protein during extrusion cooking at low and high moisture content. Excessive Maillard reaction can result in losses of lysine up to 50%, depending on barrel temperature, moisture and shear forces. Extrusion of kidney beans and chickpeas in a twin screw extruder at 150ºC did not affect the protein, ash and carbohydrate concentrations and the amino acid profiles were similar with exception of methionine and cysteine (Brenes et al. 2008, Marzo et al. 2002). Korus, Gumul, and Achremowicz (2006) studied the influence of extrusion parameter on chemical composition of seeds of five bean cultivars. They described a significant decrease (3%) in protein content of cv. Augusta extruded at 120ºC and cv. Toffi extruded at 180ºC, but an increase about 4% was noted in extrudate from cv Rawela at 120ºC, similar to the results obtained by Prakrati, Ameeta, and Kushwah (2000) in extruded faba bean. El-Hady and Habiba (2003) reported a protein content decrease about 1-5% depending on the extruded seed (pea, kidney bean, chickpea and faba bean) and an enhanced *in vitro* protein digestibility (3-11%). These authors found that significance was due to barrel temperature, while no significant effect was noticed for the moisture parameter. Poltronieri, Arêas, and Colli (2000) did not observe significant changes in the protein content of extruded chickpea var Marocos extruded at 130ºC, 14% moisture, however, Cardoso-Santiago and Arêas (2001) observed a decrease of protein content (19%) in chickpeas after extrusion.

The lipid content after extrusion was lower (50%) in pea, bean and chickpea flours in comparison to raw materials (Brenes et al. 2008, Cardoso Santiago et al. 2001, Marzo et al. 2002, Poltronieri, Arêas, and Colli 2000). This reduction would be due to the starch-lipid and protein-lipid complexes formed during extrusion that are resistant to some lipid extraction procedures. Cornmeal extruded at 85-90ºC had a 75% of its lipids bound, but extrusion at 120-125ºC bound 70% of the lipids (Guzman, Lee, and Chichester 1992). However, Korus, Gumul, and Achremowicz (2006) reported an increase about 3% in lipid content of extrudates obtained from beans cv. Nigeria, Rawela and about 7% in cv.

Tip-Top. This increase could be due to the use of a lower barrel temperature and moisture (120°C, 14%).

Starch is usually the major component in extruded food, especially in expanded products such as snacks or breakfast cereals. During extrusion process starch undergoes structural changes including loss of starch crystalline structure, destruction of granular structure, and gelatinization. The extent of these transformations depends on the extrusion parameter used (Wen, Rodis, and Wasserman 1990). Extrusion-cooking of peas at 145°C and 25% moisture content significantly increased the starch digestibility (Alonso et al. 2000). The rate of gelatinization depends on the temperature, moisture, shear and pressure used during extrusion (Camire 2002). Generally, after extrusion total carbohydrates and starch content were modified. Starch and NSP reduced in extruded pea and kidney beans (Alonso et al. 2001, Korus, Gumul, and Achremowicz 2006), but Cardoso-Santiago and Arêas (2001) described an increase about an 8% of total carbohydrates after extrusion of chickpea. Zamora (2003) observed an increase of 10% in starch content of *Canavalia ensiformis* after extrusion. Arribas, Guillamón, et al. (2016) observed an increase in the total starch content and a significant reduction in resistant starch content of different extruded rice, bean and carob bean blends at 125°C, 20% moisture. During extrusion, starch molecules are broken into smaller fragments, changing the amylose/amylopectine ratio and the proportion of resistant starch in beans and chickpeas (Escarpa et al. 1996, Poltronieri, Arêas, and Colli 2000).

Dietary fiber is an important component of foodstuff in relation to their nutritional value. The role in prevention of cardiovascular diseases, in regulation of intestine flow, in maintaining the proper blood levels, etc. During extrusion, dietary fiber may be significantly modified, both in its content and solubility. Korus, Gumul, and Achremowicz (2006) reported that the total dietary fiber content in five bean cultivars was smaller after extrusion in the range 3-9% and in all extrudates the content of insoluble fiber decreased while the soluble fiber content increases; similar result were reported by Arribas et al. (2016) after extrusion of different rice, bean and carob flour blends at 125°C, 20% moisture. The reduction was higher in samples with lower humidity (14%). Martin-Cabrejas et al. (1999) and Poltronieri, Arêas, and Colli (2000) reported that raw and extruded beans and chickpeas had similar total fiber content, however a redistribution of insoluble to soluble fiber occurred. These changes are important, since the inclusion of dietary fiber in formulations produced significant reductions in sectional expansion and increased bulk density of extruded products (Brennan, Monro, and Brennan 2008). Some formulations based on lentil, pea or chickpea flours demonstrated a beneficial increase in dietary fiber, so, the fabrication of value-added, pulse-based snacks with high dietary fiber contents would be an excellent means to provide the population with recommended amounts of this beneficial food component (Berrios et al. 2010).

In general, when we process food through thermal processing methods, as extrusion, a reasonable amount of the present vitamins is lost, although vitamin stability varies with vitamin structure, extrusion conditions, and food matrix composition (Athar et al. 2006). The most sensitive to the extrusion process are vitamin A and vitamin E from fat-soluble vitamins, and vitamin C, B_1, and folic acid from water-soluble vitamins (Riaz, Asif, and Ali 2009) (and papers cited in). Despite the increased use of extrusion, there is little information about the stability of vitamins during pulse extrusion. Stability of various forms of vitamin E varies with grass peas (*Lathyrus sativus*, L.) extrusion, so α-tocopherol content decrease with an increase in extrusion temperature, while α-tocopherol decrease with increasing moisture content during extrusion. Corn-soy blends processed at 171°C retained over 90% of thiamine, riboflavin, vitamin B_6 and folic acid (Camire, Camire, and Krumhar 1990, Riaz, Asif, and Ali 2009). Anuonye et al. (2010) observed a 6% decrease in vitamin B_2, an 86.36% decrease in vitamin B_6 and no significant change in vitamin C content. Lorenz and Jansen (1980) found that after extrusion of soy and soy-corn blends, there was no loss of thiamine during storage of these extruded products and retained 80% vitamin C.

Mineral content and bioavailability are generally retained during extrusion (Korus, Gumul, and Achremowicz 2006). Low-shear extrusion retained dialyzable iron in navy beans, lentils, chickpeas and cowpeas

(Ummadi, Chenoweth, and NG 1995). Extrusion cooking of peas and kidney beans at 150-155°C did not significantly affect the mineral composition of pea and kidney beans. In growing rats, the apparent absorption of Fe, Ca and P from pea or kidney bean-based diets significantly increased in extruded compared with raw seed meals (Alonso et al. 2001). Zinc bioavailability of semolina and soy protein concentrate blends (85/15) was unaffected by extrusion (Kang 1996).

Alonso et al. (2000) showed an increase of the protein digestibility (up to 97%) in extruded peas. Similarly, Arribas, Guillamón, et al. (2016) reported an increase in protein digestibility of extruded rice, bean and carob flour blends when extruded at 125°C, 20% moisture. A reduction in total and LDL cholesterol in rat feeds with extruded peas in comparison to a control diet was reported by Alonso et al. (2001), however, Oosthuizen et al. (2000) did not report reduction in serum lipoproteins in men with hyperlipidemia that consumed baked goods fortified with 92g/day extruded dry white beans. Brenes et al. (2008) reported that the inclusion of extruded chickpea in the diet of growing chickens improved weight gain, protein and fat digestibilities.

Many authors described the extruded products elaborated with different legumes as nutritious and healthy, however there are scarce information about the contribution of those ready-to-eat expanded products to the dietary recommendations for a balanced and healthy diet. Espinoza-Moreno et al. (2016) produced an expanded snack from whole amarantin transgenic maize and black common bean at 170°C, 25% moisture. Fifty grams portion of these snacks covers 55.6% and 38.05% of daily protein requirements for children 1-3 and 4-8 years, respectively, due to its supplementation with common bean. A portion also provides 7.5 g of total dietary fiber and contributes with 25-15% of daily requirements of antioxidants. Martín-Pedrosa et al. (2015) reported that one serving (45 g) of different extruded rice, bean and carob flour blends would provide (on average) a 11% of daily protein intake, 0.1% of the daily fat intake, 25% of the daily carbohydrate intake, 9% of the dietary fiber intake and it represents, approximately, a 7% of a 2000-kcal daily energy intake. A source of fiber must contain at least 3% of the daily food intake, while the requirements for high fiber food is 6% of the recommended daily intake, so the extruded products developed by these authors must be considered as a good source of fiber.

The nature and composition of the raw materials and the extrusion processing conditions are accounted for the quality of finished products, since they will determine the crispness, crackliness, hardness, and overall acceptability of the new products. The success or failure of a new extruded food product is directly related to sensory attributes, where texture plays a major role (Pamies et al. 2000). Due to the high cost involved in sensory analysis, instrumental assessment of end-product texture has been considered a good alternative to sensory evaluation (Anton and Luciano 2007), although usually the evaluation of extruded products is referred to a trained or untrained panel. Paula and Conti-Silva (2014) found a strong good correlation between sensory and instrumental analysis that enable application of instrumental test (compression, cut, puncture and shear tests) as an indicator of the sensory texture of extruded snacks. The sensory analysis of a chickpea-based snack developed by Batistuti, Barros, and Arêas (1991) had higher acceptance than a commercial corn snack and also presented a good digestibility. Sensory assessment of the extruded samples fried in oil indicated that extrusion of rice or rice-chickpea blends could be an alternative method of producing acceptable snacks (Bhattacharya and Choudhury 1994). Sensory evaluation of extruded products elaborated with different proportions of corn, rice, cowpea, and chickpea flours showed that the combination with higher proportions of legumes had better color, flavor, texture and overall acceptability than others (Shadan, Waghray, and Khoushabi 2014).

Thus, extruded legume-based foods were high in protein, dietary fiber, and low in fat. They present high protein and starch digestibilities, moderate amounts of vitamins and do not affect significantly the mineral content. The extrusion cooking of legumes improved their nutritional value in the same extent as some heat traditional processes allowing to develop a wide range of ready-to-eat products by this technique, and increasing the consumption of legumes.

3. Effect of Extrusion/Cooking on the Nutraceutical Value of Legumes

Legumes contain bioactive compounds, such as oligosaccharides, inositol phosphates, protease inhibitors, lectins, phenolic compounds, etc. Some of these compounds have been associated with undesirable and physiological factors that have limited the widespread use of pulses as primary foods, mainly in developed countries. However, the effect of these compounds as bioactive or anti-nutrient, mostly depends on their concentration in the food products. Different reports show that extrusion cooking reduces many toxins and anti-nutrients presents in legumes, but it is not desirable to abolish these compounds since in small quantities, they may impair health benefits against chronic diseases. The extrusion conditions have the ability to produce both positive or negative influences on the bioactive compounds content of the extrudates (Brennan et al. 2011).

Extrusion cooking at 150°C reduced phytic acid, trypsin, chymotrypsin and α-amylase inhibitors, condensed tannins and polyphenols of the pea and kidney bean seeds. Furthermore, hemagglutinating activity was completely abolished. (Alonso et al. 2001; Marzo et al. 2002). Reductions in protease inhibitors, lectins, tannins and phytates can contribute to better legume protein utilization, since these compounds reduced protein digestibility (Alonso, Orue, and Marzo 2000). Inositol phosphates play critical roles in diseases such as cancer, diabetes type 2, irritable bowel syndrome (Shi et al. 2006), while trypsin inhibitors were effective in preventing or suppressing carcinogen-induced effects (Clemente et al. 2004). The phytate and tannin reduction can improve the mineral bioavailability in extruded foods. Morales et al. (2015) reported an overall reduction in the total inositol phosphates content (14-50%) in different extruded lentil flours at 160°C. They observed a significant increase in IP5 and IP4 forms. Similarly, Butron (2002) reported a decrease of inositol phosphates content in faba bean (13.7%) and in kidney bean (22%) processed at 160°C. However, a lower reduction in phytic content (2-9%)

of extruded faba bean, chickpeas, peas and kidney beans was reported by El-Hady and Habiba (2003) and they concluded that the higher temperature used (180°C) was significantly more effective in reducing phytic acid than the lower one (140°C). Varela et al. (2007) also reported a decrease of inositol phosphates in different blends of pulses/corn starch (80/20); the reduction was around an 8% in kidney bean blends, an 8.5% in chickpea blends and a 35.8% in faba bean blends after extrusion cooking of the pulse/corn starch flours at 150°C. Anton, Gary Fulcher, and Arntfield (2009) observed an overall reduction of nearly 44% after extrusion of navy/small red bean and corn starch blends. On the contrary, Lombardi-Boccia, Lullo, and Carnovale (1991) reported that legume phytate content was not affect by extrusion. The reduction produced in phytate (IP6) content may be due in part to hydrolysis of the phytic acid into lower phosphate forms (IP5 to IP1) that are considered to have a very important role in human health and diseases (Alonso, Orúe, and Marzo 2000, Harland and Morris 1995, Morales et al. 2015). So, extrusion could be a good alternative to reduce phytic acid (IP6) content and maintain beneficial, less phosphorylated inositol phosphates, in the extrudates.

Trypsin inhibitors are heat-labile compounds, so extrusion cooking reduced significantly their content. The extent of the reduction varies with the seeds, the formulation flours and the parameters of the extrusion process. Zamora (2003) reported a reduction about 95% of trypsin and chymotrypsin inhibitor activities in extrudates of *Canavalia ensiformis*. Alonso, Orúe, and Marzo (2000) found differences in the reduction on trypsin and chymotrypsin inhibitors in extruded peas and kidney beans, so trypsin inhibitory activity was reduced about a 95% but chymotrypsin inhibitory activity was also reduced (65%) but not so extensively. This fact could be related to the presence of different isoforms of the inhibitor with different heat sensitivity. According to Varela et al. (2007) extrusion at 150°C of pea, kidney bean, chickpea, faba bean or lupin/starch blends almost abolished the activity of the trypsin inhibitors except for chickpeas which showed a reduction about 90%. The residual activity would be probably due to this chickpea isoforms are more resistant to extrusion treatment than that from others legumes. These changes were confirmed by zymograms gels; four different isoforms of trypsin inhibitor were clearly detected in raw peas and kidney beans, two in chickpeas and only one in faba beans. Except in kidney beans, the inhibitors appeared in zones of higher electronegativity than the soybean inhibitor used as a standard. After extrusion no bands could be detected in any case. Extrusion decreased the trypsin inhibitor content in different lentil-based formulations, although the extrusion processing did not affect all formulations to the same extent. The greatest losses (97%) were observed in the extruded samples formulated with lentil plus apple fiber and/or corn flour and the lower reductions (93%) corresponded to formulation containing wheat bran (Morales et al. 2015). Anton et al. (2009) reported that extrusion at 160°C, 22% moisture was able to reduce trypsin inhibitors by 100% in all samples. Similarly, El-Hady and Habiba (2003) reported that extrusion of peas, chickpeas, faba and kidney bean at 140 or 180°C completely abolished the trypsin inhibitor activity.

Lectins are proteins or glycoproteins with a high resistance to proteolytic *in vivo* degradation and their ability to recognize and bind to sugar moieties of intestinal epithelial cells may result in hyperplasia. Lectins are considered as the main anti-nutrient component in beans. However, a small amount of lectins may be beneficial in stimulating gut function, limiting tumor growth and ameliorating obesity. As observed for protease inhibitors, extrusion processing at temperatures higher than 145°C abolish the lectin content in pulses (lentil, pea, bean, faba bean, chickpea, canavalia) demonstrating the heat sensitive nature of this compound (Alonso, Aguirre, and Marzo 2000, Armour et al. 1998, Arribas, Belmonte, et al. 2016, Espinoza-Moreno et al. 2016, Leontowicz et al. 1999, Morales et al. 2015, Varela et al. 2007, Zamora 2003). Although, Kelkar et al. (2012) observed a small reduction (about 90%) in PHA activity after extrusion of navy and pinto beans at 85°C and 36% moisture.

The raffinose family of oligosaccharides or α-galactosides (flatulence factors) have been identified as prebiotic agents, they are fermented by colonic flora to produce a mixture of short-chain fatty acids (SCFA) that stimulated bifidobacterias growth. This fact may have several beneficial implications for health: potential protective effects against colorectal cancer and infectious bowel diseases. Moreover, the SCFA could also be responsible for lowering cholesterol and glycemic symptoms after being absorbed in the colon (Lajolo et al. 2004). Although, the effect of extrusion on the content of soluble sugars is controversial; some authors reported a reduction, while other authors reported an increase on the sugar content of the extruded pulses. The differences could be a consequence of the conditions used: temperature, screw speed or moisture. Berrios and Pan (2001) and Berrios et al. (2002) reported that extrusion processing reduced the total oligosaccharide levels in black bean flours and that the extent of their reduction depended on the extrusion conditions employed. The changes in the α-galactosides content has been reported that also depend on the seed type. Varela et al. (2007) studying the changes in total α-galactosides content in different extruded pulses/corn starch blends at 150°C and 20% moisture, observed that not all the samples are affected in the same extent, so extruded pea flours showed an increase about 15.18%, whereas a decrease in the sugar content was found for the other extruded legumes. The highest reduction was found in extruded faba bean seeds (47%), and the lowest one corresponded to extruded kidney beans (3%). Sucrose and α-galactosides (raffinose and stachyose) decreased significantly in extruded pinto bean with higher reduction at higher process temperature (110-163°C) (Borejszo and Khan 1992). Raffinose, stachyose and verbascose dropped in kidney bean after extrusion at 150°C and 20% moisture, but only stachyose was reduced in extruded pea flours compared with raw seeds (Alonso, Aguirre, and Marzo 2000). Berrios et al. (2010) studied the carbohydrate composition of extruded lentil, chickpea and pea flours. The highest carbohydrate content corresponded to pea formulations, followed of chickpea and lentil. Extrusion at 160°C and 17% moisture did not significantly affect the total available carbohydrates, but, a reduction in the concentration of the α-galactosides in lentil and pea extrudates was observed, however, chickpea extrudates showed a higher concentration of raffinose, stachyose and ciceritol. These authors concluded that these pulse-based snacks with reduced levels of flatulence factors and higher contents of dietary fiber represent good alternatives to traditional cereal-based snacks. Morales et al. (2015) studied the effect of extrusion cooking on different gluten-free snacks elaborated with lentil flours and formulated fiber-rich lentil flours. The extrusion was developed at 160°C and 17% moisture. The extruded flours presented a higher sucrose and galactosides content than the unprocessed flours. This could be attributed to the hydrolysis of some soluble sugars during the extrusion processing and also to the mechanical-structure modifications of the product matrix, including cell wall breakage, increased porosity, improving the diffusion solvent inside the matrix, and subsequently increased the extraction of the soluble sugars as occurred with other high-pressure techniques (Amor et al. 2008, Pedrosa et al. 2012). Taking into account the health benefits associated to α-galactosides, the increase in their availability, generated as a consequence of extrusion processing, may be considered as an added-value ingredient with potential in functional food.

Phenolic compounds are the focus of numerous studies to elucidate their role in human health. They are potent antioxidant and the may help prevent some cancer. Depending on the plant material, the extrusion conditions and the phenolic compound studied, some authors reported increases or decreases in the phenols content and in the total antioxidant activity in different pulses. Korus, Gumul, and Czechowska (2007) reported that the effect of extrusion on the total phenolic content of beans depended on the cultivar, since one cultivar showed a 14% increase in the amount of phenols, while other two cultivars exhibited a decrease by 19 and 21% compared to raw beans. Korus et al. (2007) examined the effect of extrusion parameter, and they observed the highest decrease on polyphenol content and antioxidant activity of extruded common bean at 180°C and 20% moisture, while the least adverse effect was obtained at 120°C and 20% moisture. Similarly, Delgado-Licon et al. (2009) examine the influence of extrusion parameter on the phenolic compounds and the antioxidant capacity of bean/corn mixtures (60/40). Extrusion was performed in four different moisture (14.5-18%) and temperature (150°-180°C) conditions. They observed a significant decrease in total polyphenols, flavonoids and antioxidant activity in all extruded samples, but the extent of the changes was dependent on process conditions, being relatively low moisture (16.3%) and temperature (142°C) the conditions that retain higher content of polyphenols, flavonoids and antioxidant activity. El-Hady and Habiba (2003) observed a significant decrease in total phenols content at different extrusion conditions (140°C and 180°C) and moisture (18% and 22%) for pea, chickpea, faba and kidney beans. Anton, Gary Fulcher, and Arntfield (2009) described a significant decrease in the total phenols content and in the antioxidant capacity of extruded snacks containing navy/small red bean flours and corn starch. The reduction was higher in the snacks made with small red bean cultivar (70%) compared to navy bean (10%). So, the effect of bean cultivar was highly relevant on the nutraceutical profile of the extrudates. It is known that the antioxidant activity in common bean are mainly due to the presence of polyphenolic compounds which are responsible for bean seed coat color. Nayak et al. (2011) observed an increase in the total phenolic content, flavonoids and antioxidant activity in extruded products of potato and pea flour mixtures. Guillamón et al. (2016) found a slight increase in the total phenols content of different extruded formulation based on rice, bean or pea and carob flours. These authors also described a significant positive correlation between the antioxidant activity and the total phenolic content. Limsangouan et al. (2010) studied the functional properties of cereal (rice, maize) and legume (red kidney bean, broad bean, soy and pea) based extruded snack foods fortified with by-products from herbs (green tea and egoma leaves) and vegetables (carrot, garlic). They reported that extrusion at 150°C slightly decreased the antioxidant capacity and phenolic content, but substantially decreased the resistant starch content for all extruded products assayed, so the extrusion process was suitable to produce functional foods, while retained their antioxidant capacity. So, extrusion cooking of pulses is advantageous to obtain snacks or other food products with higher levels of phenolic compounds and good sensorial acceptance.

CONCLUSION

Extruded products are of high interest for the food industry since extrusion is a highly versatile process that can be applied to a variety of foods for producing new and creative products with high consumer's acceptance. Legumes are important foods because of their nutritional and bioactive compounds and their consumption provides several health benefits, so, legumes must be considered as functional foods. Legumes are processed prior to consumption and it affects the composition of macronutrient and bioactive compounds. It has been reported that depending on the concentration, the phytochemical present in the grain legumes could act as an anti-nutrient or as a bioactive compound with a health-promoting function, so it is important to know the amount of each component present in the legume-based foods that are able to exert a healthy function.

Effect of extrusion on the chemical composition of legumes depends on the raw material types, their composition and process parameters. The processing parameters arc also accounted for the quality of finished products. In general, extrusion improves protein and starch digestibilities, increased soluble dietary fiber, reduced the content of some vitamins, maintain the mineral content and reduce or eliminate some anti-nutritional factors. Although, it is important to notice that the amount of bioactive components retained in the extruded product would be enough to produce legume-based foods with improved functional functions.

So, "ready-to-eat" expanded legume-based products are recommended to improve the nutritive and functional value of legumes, and would help to expand markets and increase its utilization in human nutrition by direct consumption or as an ingredient of fortified foods.

REFERENCES

Alam, M. S., Jasmeen Kaur, Harjot Khaira, and Kalika Gupta. 2016. "Extrusion and Extruded Products: Changes in Quality Attributes as

Affected by Extrusion Process Parameters: A Review." *Critical Reviews in Food Science and Nutrition* 56 (3):445-473.

Alonso, R, A. Aguirre, and F. Marzo. 2000. "Effect of extrusion and traditional processing methods on antinutrients and in vitro digestibility of protein and starch in faba and kidney beans." *Food Chemistry* 68:159-165.

Alonso, R., L. A. Rubio, M. Muzquiz, and F. Marzo. 2001. "The effect of extrusion cooking on mineral bioavailability in pea and kidney bean seed meals." *Animal feed Science and Technology* 94:1-13.

Alonso, R., G. Grant, P. Dewey, and F. Marzo. 2000. "Nutritional Assessment *in vitro* and *in vivo* of raw and extruded peas (*Pisum sativum* L.)." *Journal Agricultural and Food Chemistry* 48:2286-2290

Alonso, R., E. Orúe, and F. Marzo. 2000. "Effect of extrusion cooking on structure and functional properties of pea and kidney bean proteins." *Journal Science Food Chemistry* 80:397-403.

Alonso, R., E. Orúe, and F. Marzo. 1998. "Effects of extrusion and conventional processing methods on protein and antinutritional factor contents in pea seeds." *Food Chemistry* 63 (4):505-512.

Allen, Karin E., Charles E. Carpenter, and Marie K. Walsh. 2007. "Influence of protein level and starch type on an extrusion-expanded whey product." *International Journal of Food Science & Technology* 42 (8):953-960.

Amor, B. B., C. Lamy, P. Andre, and K. Allaf. 2008. "Effect of instant controlled pressure drop treatments on the oligosaccharides extractability and microstructure of Tephrosia purpurea seeds." *Journal of Chromatography A,* 1213:118-124.

Anton, A. A., and F. B. Luciano. 2007. "Instrumental texture evaluation of extruded snack foods: A review evaluación instrumental de textura en alimentos extruidos: Una revisión." *Ciencia y Tecnologia Alimentaria* 5 (4):245-251.

Anton, Alex A., R. Gary Fulcher, and Susan D. Arntfield. 2009. "Physical and nutritional impact of fortification of corn starch-based extruded snacks with common bean (*Phaseolus vulgaris* L.) flour: Effects of bean addition and extrusion cooking." *Food Chemistry* 113 (4):989-996.

Anuonye, Julian C., John O. Onuh, Evans Egwim, and Samuel O. Adeyemo. 2010. "Nutrient and antinutrient composition of extruded acha/soybean blends." *Journal of Food Processing and Preservation* 34:680-691.

Armour, Julia C., R. L. Chanaka Perera, Wendy C. Buchan, and George Grant. 1998. "Protease inhibitors and lectins in soya beans and effects of aqueous heat-treatment." *Journal of the Science of Food and Agriculture* 78 (2):225-231.

Arribas, C., J. M. Belmonte, E. Guillamón, C. Cuadrado, and M.M. Pedrosa. 2016. "Changes on the lectin content of different extruded rice/legume composite flours." The Food Factor I Conference, Barcelona (Spain).

Arribas, C., E. Guillamón, C. Burbano, M. Muzquiz, C. Cuadrado, B. Cabellos, and M. M. Pedrosa. 2016. "Nutritional and nutraceutical characterization of extruded gluten-free snacks made from rice, bean and carob bean flour mixtures." Second International Legume Society Conference: Legumes for a Sustainable World, Troia (Portugal).

Azarnia, Sorayya, Joyce I. Boye, Tom Warkentin, and Linda Malcolmson. 2011. "Changes in volatile flavour compounds in field pea cultivars as affected by storage conditions." *International Journal of Food Science and Technology* 46 (11):2408-2419.

Batistuti, José Paschoal, Rosa Maria Cerdeira Barros, and José Alfredo Gomes Arêas. 1991. "Optimization of Extrusion Cooking Process for Chickpea (Cicer arietinum, L.) Defatted Flour by Response Surface Methodology." *Journal of Food Science* 56 (6):1695-1698.

Berrios, J. De J., M. Cámara, M. E. Torija, and M. Alonso. 2002. "Effect of extrusion cooking and sodium bicarbonate addition on the carbohydrate composition of black bean flours." *Journal of Food Processing and Preservation* 26 (2):113-P128.

Berrios, J., De J., P. Morales, M. Cámara, and M. C. Sánchez-Mata. 2010. "Carbohydrate composition of raw and extruded pulse flours." *Food research International* 40 (531-536).

Berrios, J. D. 2006. "Extrusion Cooking of Legumes: Dry Bean Flours." *Encyclopedia of Agricultural, Food, and Biological Engineering.* 1:1-8.

Berrios, J. De J., and J. Pan. 2001. "Evaluation of extruded black bean (Phaseolus vulgaris L.) processed under different screw speeds and particle sizes." Annual meeting of the Institute of Food technologist, New Orleans, LA.

Bhattacharya, S., and G.S. Choudhury. 1994. "Twin-screw extrusion of rice flour: Effect of extruder length-to-diameter ratio and barrel temperature on extrusion parameters and product characteristics." *Journal of Food Process and Preservation* 18 (5):389-406.

Borejszo, Z, and k. Khan. 1992. "Reduction of flatulence-causing sugars by high temperature extrusion of pinto bean high starch fractions." *Journal of Food Science* 57 (3):771-777.

Boye, Joyce, Fatemeh Zare, and Alison Pletch. 2010. "Pulse proteins: Processing, characterization, functional properties and applications in food and feed." *Food Research International* 43 (2):414-431.

Brenes, A., A. Viveros, C. Centeno, I Arija, and F. Marzo. 2008. "Nutritional value of raw and extruded chickpeas (*Cicer arietinum* L.) for growing chickens." *Spanish Journal of Agricultural Research* 6 (4):537-545.

Brennan, Charles, Margaret Brennan, Emma Derbyshire, and Brijesh K. Tiwari. 2011. "Effects of extrusion on the polyphenols, vitamins and antioxidant activity of foods." *Trends in Food Science & Technology* 22 (10):570-575.

Brennan, Margaret A., John A. Monro, and Charles S. Brennan. 2008. "Effect of inclusion of soluble and insoluble fibres into extruded breakfast cereal products made with reverse screw configuration." *International Journal of Food Science & Technology* 43 (12):2278-2288.

Butron, J. 2002. Thesis: "Effect of extrusion of legumes and enzyme supplementation on the animal growth and their metabolism." *Biological Science,* Universidad Pública de Navarra (Spain).

Camire, M.E. 2002. "Extrusion cooking." In *The Nutrition Handbook for Food Processors*, edited by C. J. K. Henry and C. Chapman, 314-330. Boca Raton, FL. USA: CRC Press.

Camire, Mary Ellen, Alex Camire, and Kim Krumhar. 1990. "Chemical and nutritional changes in foods during extrusion." *Critical Reviews in Food Science and Nutrition* 29 (1):35-57.

Campos-Vega, R, G Loarca-Piña, and B.D Oomah. 2010. "Minor components of pulses and their potential impact on human health." *Food Research International* 43:461-482.

Cardoso-Santiago, R. A., and J. A. G. Arêas. 2001. "Nutritional evaluation of snacks obtained from chickpea and bovine lung blends." *Food Chemistry* 74 (1):35-40.

Cardoso Santiago, R. A., R. S. R. Moreira-Araújo, M. E. M. Pinto e Silva, and J. A. G. Arêas. 2001. "The potential of extruded chickpea, corn and bovine lung for malnutrition programs." *Innovative Food Science & Emerging Technologies* 2 (3):203-209.

Castells, M., S. Marín, V. Sanchis, and A. J. Ramos. 2005. "Fate of mycotoxins in cereals during extrusion cooking: A review." *Food Additives and Contaminants* 22 (2):150-157.

Clemente, A., D. A. Mackenzie, I. T. Johnson, and C. Domoney. 2004. "Investigation of legume seed protease inhibitors as potential anticarcinogenic proteins." In *Recent Advances of Research in Antinutritional Factors in Legume Seeds and Oilseeds.*, edited by G.D. Hill, M. Muzquiz, C. Cuadrado, M.M. Pedrosa, and C. Burbano, 137-142. Toledo (Spain): EAAP publication.

Chen, Feng Liang, Yi Min Wei, and Bo Zhang. 2011. "Chemical cross-linking and molecular aggregation of soybean protein during extrusion cooking at low and high moisture content." *LWT - Food Science and Technology* 44 (4):957-962.

DAFNE. 2009. "Pan-European Food Data Bank based on Household Budget Surveys.". Data Food Networking, accessed June. <http://nut.uoa.gr/Dafnesoftweb>.

Delgado-Licon, E., A. L. Ayala, N E. Rocha Guzman, J. A. Gallegos-Infante, M. Atienzo-Lazos, J. Drzewiecki, C. E. Martinez-Sanchez, and S. Gorinstein. 2009. "Influence of extrusion on the bioactive compounds and the antioxidant capacity of the bean/corn mixtures." *International Journal Food Science Nutrition* 60 (6):522-32.

El-Hady, E. A., and R. A. Habiba. 2003. "Effect of soaking and extrusion conditions on antinutrients and protein digestibility of legume seeds." *LWT - Food Science and Technology* 36 (3):285-293.

Escarpa, A., M. C. González, E. Mañas, L. García-Diz, and F. Saura-Calixto. 1996. "Resistant starch formation: standardization of a high-pressure autoclave process." *Journal of Agricultural and Food Chemistry* 44 (3):924-928.

Espinoza-Moreno, Ramona J., Cuauhtémoc Reyes-Moreno, Jorge Milán-Carrillo, José A. López-Valenzuela, Octavio Paredes-López, and Roberto Gutiérrez-Dorado. 2016. "Healthy ready-to-eat expanded snack with high nutritional and antioxidant value produced from whole amarantin transgenic maize and black common bean." *Plant Foods for Human Nutrition* 71:218–224.

FAO. 2004. Definition and classification of commodities, pulses and derived products.

FAOSTAT. 2016. "Food And Agricultural Organization of United Nations: Economic and Social Department: The Statistical Division." http://fao.org/faostat/en/3search/pulses%2BTotal (accessed October 2016).

Guillamón, E., C. Arribas, C. Burbano, M. Muzquiz, C. Cuadrado, B. Cabellos, and M.M. Pedrosa. 2016. "Phenolic content and antioxidant activity in gluten-free extruded composite flours of rice and different legumes" Second International legume Society Conference: Legumes for a Sustainable World, Troia (Portugal).

Guzman, L. B., T. C. Lee, and C. 0. Chichester. 1992. "Lipid binding during extrusion cooking." In *Food Extrusion Science and Technology,* edited by H. Chi-Tang J. L. Kokini, and M. V. Karwe, 693-709. Marcel Dekker: New York.

Harland, BF, and ER Morris. 1995. "Phytate: a good or a bad component?" *Nutrition Research* 15:733-754.

Harper, J. M. 1989. "Food extruders and their applications." In *Extrusion Cooking*, edited by C. Mercier, P. Linko and J.M. Harper, 1-16. St Paul, MN: American Association of cereal Chemists, Inc.

Kang, S. Y. 1996. Thesis. "Zinc bioavailability in a semolina/soy protein mixture was not affected by extrusion processing." East Lansing, MI, Michigan State University.

Kelkar, S., M. Siddiq, J. B. Harte, K. D. Dolan, G. Nyombaire, and H. Suniaga. 2012. "Use of low-temperature extrusion for reducing phytohemagglutinin activity (PHA) and oligosaccharides in beans (*Phaseolus vulgaris* L.) cv. Navy and Pinto." *Food Chemistry* 133 (4):1636-1639.

Korus, J., D. Gumul, and B. Achremowicz. 2006. The influence of extrusion on chemical composition of dry seeds of bean (*Phaseolus vulgaris* L.). *Electronic Journal of Polish Agricultural Universities* 9 (1).

Korus, J., D. Gumul, and K. Czechowska. 2007. "Effect of extrusion on the phenolic composition and antioxidant activity of dry beans of *Phaseolus vulgaris* L." *Food technology Biotechnology* 45 (2):139-146.

Korus, J., D. Gumul, M. Folta, and H. Barton. 2007. Antioxidant and antiradical activity of raw and extruded commons beans. *Electronic Journal of Polish Agricultural Universities* 10 (4).

Lajolo, F. M., M. I. Genovese, I. F. Pryme, and T. M. Dale. 2004. "Beneficial (antiproliferative) effects of different substances." In *Recent advances of research in antinutritional factors in legume seeds and oilseeds*, edited by G.D. Hill, M. Muzquiz, C. Cuadrado, M.M. Pedrosa, and C. Burbano, 123-137. Toledo (Spain): EAAP publication.

Leontowicz, H., M. Leontowicz, H. Kostyra, Gralak M.A., and G.W. Kulasek. 1999. "The influence of extrusion or boiling on trypsin inhibitor and lectin activity in leguminous seeds and protein digestibility in rats." *Polish Journal of Food and Nutrition Sciences* 8:77-87.

Limsangouan, Nipat, Makiko Takenaka, Itaru Sotome, Nanayama Kazuko, Charunuch Chulaluck, and Isobe Seiichiro. 2010. "Functional properties of cereal and legume based extruded snack foods fortified with by-products from herbs and vegetables." *Kasetsart J. (Nat. Sci.)* 44:271-279.

Lombardi-Boccia, Ginevra, Giuseppe Di Lullo, and Emilia Carnovale. 1991. "*In-vitro* iron dialysability from legumes: Influence of phytate and extrusion cooking." *Journal of the Science of Food and Agriculture* 55 (4):599-605.

Lorenz, K., and G. R. Jansen. 1980. "Nutrient stability of full-fat soy flour and corn-soy blends produced by low-cost extrusion." *Cereal Foods World* 25 (4):161-172.

Martin-Cabrejas, M. A., L. Jamie, C Karanja, A J. Downie, M. L. Parker, F. J. Lopez-Andreu, G. Maina, R. M. Esteban, A. C. Smith, and K. W. Waldron. 1999. "Modifications to physicochemical and nutritional properties of hard-to-cook beans (*Phaseolus vulgaris* L) by extrusion cooking." *Journal Agricultural Food Chemistry* 47 (3):1174-1182.

Martín-Pedrosa, M, C. Arribas, B. Cabelos, C Cuadrado, C Burbano, M Muzquiz, and E Guillamón. 2015. "Development and characterization of rice based extruded product foods fortified with legume flours." Eucarpia International Symposium on Protein crops. V Meeting AEL, Pontevedra (Spain).

Marzo, F., R. Alonso, E. Urdaneta, F. J. Arricibita, and F. Ibanez. 2002. "Nutritional quality of extruded kidney bean (*Phaseolus vulgaris* L. var. Pinto) and its effects on growth and skeletal muscle nitrogen fractions in rats." *Journal Animal Science* 80 (4):875-9.

Mollet, B., and I. Rowland. 2002. "Functional foods: at the frontier between food and pharma." *Current Opinion in Biotechnology* 13:483-485.

Morales, P., J. De J. Berrios, A. Varela, C. Burbano, C. Cuadrado, M. Muzquiz, and M. M. Pedrosa. 2015. "Novel fiber-rich lentil flours as snack-type functional foods: an extrusion cooking effect on bioactive compounds." *Food and Function* 6:3135-3143.

Nayak, Balunkeswar, Rui Hai Liu, Jose De J. Berrios, Juming Tang, and Christopher Derito. 2011. "Bioactivity of antioxidants in extruded products prepared from purple potato and dry pea flours." *Journal of Agricultural and Food Chemistry* 59 (15):8233-8243.

Oosthuizen, W., C. S. Scholtz, H. H. Vorster, J. C. Jerling, and W. J. H. Vermaak. 2000. "Extruded dry beans and serum lipoprotein and plasma haemostatic factors in hyperlipidaemic men." *European Journal Clinical Nutrition* 54 (5):373-379.

Pamies, B. Valles, G. Roudaut, C. Dacremont, M. Le Meste, and JR Mitchell. 2000. "Understanding the texture of low moisture cereal products: mechanical and sensory measurements of crispness." *Journal of the Science of Food and Agriculture* 80 (11):1679-1685.

Paula, Amanda Maldo, and Ana Carolina Conti-Silva. 2014. "Texture profile and correlation between sensory and instrumental analyses on extruded snacks." *Journal of Food Engineering* 121:9-14.

Pedrosa, MM, C. Cuadrado, C. Burbano, K. Allaf, J Haddad, E. Gelencser, K. Takács, E. Guillamón, and M. Muzquiz. 2012. "Effect of instant controlled pressure drop on the oligosaccharides, inositol phosphates, trypsin inhibitors and lectins content of different legumes." *Food Chemistry* 131:862-868.

Poelman, A. A. M., C. M. Delahunty, and C. de Graaf. 2013. "Cooking time but not cooking method affects children's acceptance of Brassica vegetables." *Food Quality and Preference* 28 (2):441-448.

Poltronieri, F., J. A. G. Arêas, and C. Colli. 2000. "Extrusion and iron bioavailability in chickpea (*Cicer arietinum* L.)." *Food Chemistry* 70 (2):175-180.

Prakrati, Rajawat, Kushwah Ameeta, and H. S. Kushwah. 2000. "Effect of extrusion cooking variables on biochemical composition of faba bean (Vicia faba L.)." *Journal of Food Science and Technology (Mysore)* 37 (4):373-379.

Riaz, M., M. Asif, and R. Ali. 2009. "Stability of vitamins during extrusion." *Critical Reviews in Food Science and Nutrition* 49:361-368.

Shadan, Mohammad Reza, Kavita Waghray, and Fahimeh Khoushabi. 2014. "Formulation, preparation and evaluation of low-cost extrude products based on cereals and pulses." *Food and Nutrition Sciences* 5:1333-1340.

Shi, Yihui, Abed N. Azab, Morgan N. Thompson, and Miriam L. Greenberg 2006. "Inositol phosphates and phosphoinositides in health and disease." In *Biology of Inositols and Phosphoinositides*, edited by A. Lahiri Majumder and B. B. Biswas, 265-292. Springer US.

Sievenpiper, J. L., C. W. Kendall, A. Esfahani, J. M. Wong, A. J. Carleton, H. Y. Jiang, R. P. Bazinet, E. Vidgen, and D. J. Jenkins. 2009. "Effect of non-oil-seed pulses on glycaemic control: a systematic review and meta-analysis of randomised controlled experimental trials in people with and without diabetes." *Diabetologia* 52 (8):1479-95.

Singh, Shivendra, Shirani Gamlath, and Lara Wakeling. 2007. "Nutritional aspects of food extrusion: a review." *International Journal of Food Science & Technology* 42 (8):916-929.

Stojceska, Valentina, Paul Ainsworth, Andrew Plunkett, and Şenol İbanoğlu. 2009. "The effect of extrusion cooking using different water feed rates on the quality of ready-to-eat snacks made from food by-products." *Food Chemistry* 114 (1):226-232.

Ummadi, P., W.L. Chenoweth, and P.K.W. NG. 1995. "Changes in solubility and distribution of semolina proteins due to extrusion processing." *Cereal Chemistry* 72 (6):564-567.

US Dietary, Guidelines for Americans. 2010. U.S. Department of Health and Human Services.

Varela, A., E. Guillamón, C. Cuadrado, F. Marzo, C. Burbano, M. Muzquiz, and M. M. Pedrosa. 2007. "Changes in nutritional active factors and protein in different legumes by extrusion/cooking." 6th European Conference on Grain Legumes, Lisboa (Portugal).

Wen, L-F., P. Rodis, and B. P. Wasserman. 1990. "Starch fragmentation and protein insolubilization during twin-screw extrusion of corn meal." *Cereal Chemistry* 67:268-275.

Zamora, N. C. 2003. "Effect of extrusion on the activity of anti-nutritional factors and in vitro digestibility of protein and starch in flours of *Canavalia ensiformis*." *Archives Latinoamican Nutrition* 53 (3):293-8.

In: Legumes for Global Food Security ISBN: 978-1-53612-265-7
Editors: Alfonso Clemente et al.

Chapter 8

FORAGE LEGUMES WITH IMPROVED NUTRITIONAL VALUE: CONDENSED TANNINS TO AVOID PASTURE BLOAT

Luis A. Cañas, Sandra Fresquet, Edelín M. Roque, M. Cruz Rochina and José Pío Beltrán*
Department of Biology & Biotechnology of Reproductive Development, Instituto de Biología Molecular y Celular de Plantas (CSIC-UPV), Valencia, Spain

ABSTRACT

Global food security is a challenge we have to reach in the present century. In this context, the production of more safe forages to feed livestock is a main objective of most research projects worldwide. Pasture bloat is a digestive pathology producing yearly a great number of deaths among the ruminant livestock. Bloat is the result of an accumulation of gases in the rumen as result of microbial fermentations that the animal is unable to expel by eructation. It has been proved that the presence in the forage of condensed tannins or proanthocyanidins (PAs) considerably

* Corresponding Author Email: jbeltran@ibmcp.upv.es.

reduces the presence of greenhouse gases (CO_2, CH_4) in the rumen, avoiding foam formation and gas accumulation. Our group is currently working in a project to *de novo* activate the anthocyanin and proanthocyanidin pathways in alfalfa, a forage legume with high protein content but lacking PAs. To achieve this objective we have developed two biotechnological tools: an efficient Agrobacterium-mediated genetic transformation protocol via somatic embryogenesis of different alfalfa cultivars, and a multigenic construct containing two transcription factors from *Antirrhinum majus* (*AmRosea1* and *AmDelila*) to activate the anthocyanin pathway and produce the precursors of PAs, and two *Medicago truncatula* genes (*MtANR* and *MtLAR*) codifying the two enzymes necessary to activate the PAs pathway. Preliminary results obtained to *de novo* activate both pathways are promising because this construct has been successfully validated in a model system through stable transformation of *Arabidopsis thaliana* plants. In the transgenic Arabidopsis plants we have corroborated the expression of the four transgenes and the activation of the anthocyanin pathway. Stable transformation assays of different alfalfa genotypes with our multigenic construct in combination with the genetic transformation method developed are currently in progress to produce transgenic alfalfa plants with improved nutritional value.

Keywords: pasture bloat, alfalfa, condensed tannins, genetic transformation, metabolic engineering

1. Introduction

Ruminant livestock (cattle, goats, sheep, etc.) are able to acquire nutrients from plant-based food by fermenting it in a specialized compartment (rumen) prior to digestion, principally through microbial action. The process typically requires the fermented ingesta to be regurgitated and chewed again. The process of rechewing the cud to further break down plant matter and stimulate digestion is called rumination. Ruminants have a complicated polygastric stomach with four-compartments: rumen, reticulum, omasum and abomasum (Figure 1A). In the first two chambers the food is mixed with saliva and separates into layers of solid and liquid material. Solids clump together to form the cud, which is then regurgitated and chewed to completely mix it with saliva and to break down the particle size. Fiber, especially cellulose and hemicellulose, is primarily broken down in these chambers by microbes into the three volatile fatty acids (acetic, propionic and butyric). Protein and non-structural carbohydrate (pectin, sugars and starches) are also fermented. The reticulum allows the animal to regurgitate and reprocess particulate matter. More finely-divided food is then passed to the omasum, for further mechanical processing. The mass is finally passed to the true stomach, the abomassum, where the digestive enzyme lysozyme breaks down the bacteria so as to release nutrients before entering the duodenum. Use of plant material is thus indirect, with primary processing by the bacterial flora maintained in the stomach. Normal digestive processes in ruminant livestock produce greenhouse gases consisting chiefly of carbon dioxide and methane in the rumen. Under normal conditions, the gas produced in the rumen separates from the solid and liquid contents and then rises to the top of the rumen, where it collects as a free bubble. Eructation, or belching, is initiated by increasing gas pressure in the rumen. When an animal belches, the rumen contracts and pushes the free gas toward the front of the rumen, where it collects around the esophageal opening.

1.1. Pasture Bloat: A Big Problem for Ruminant Livestock Producers

Pasture bloat could be considered of a severe indigestion marked by an accumulation of gases in the rumen that the animal is unable to expel by eructation. Gases that are trapped may form a foam or froth in the rumen which further prevents their elimination. Foam formation can be caused by many factors resulting from interactions between the animal, rumen microbiota and differences in plant biochemistry. Visual signs of bloated cattle include distension of the left side of the abdomen associated with dyspnea (Figure 1B), discomfort as indicated by stomping of feet or kicking of belly, laboured breathing, frequent urination and defecation, and sudden collapse. Many cattle die by asphyxiation unless the bloat is not resolved by eructation or by manual or surgical alleviation of the gas or fluid accumulated. Death rates as high as 20% are recorded in cattle grazing bloat-prone pasture, and in pastoral areas, the annual mortality rate from bloat in dairy cows may approach 1%. There is also economic loss from depressed milk production in nonfatal cases and from suboptimal use of bloat-prone pastures. Bloat can be a significant cause of mortality in feedlot cattle (Jonker and Yu, 2016). In life-threatening cases, a trocar and cannula may be used for emergency relief (Figure 1C) although the standard-sized instrument is not large enough to allow the viscous, stable foam in peracute cases to escape quickly enough (Figure 1D). A larger bore instrument (2.5 cm in diameter) is necessary, but an incision through the skin must be made before it can be inserted through the muscle layers and into the rumen. If the cannula provides some relief, an antifoaming agent can be administered through the cannula, which can remain in place until the animal has returned to normal, usually within several hours. If the cannula fails to reduce the bloat and the animal's life is threatened, an emergency rumenotomy should be performed.

In primary ruminal tympany (frothy bloat) the cause is entrapment of the fermentation gases in stable foam. Coalescence of the small gas bubbles is inhibited, and intraruminal pressure increases because eructation cannot occur. Several factors, both animal and plant, influence the formation of a stable foam. Soluble leaf proteins, saponins, and hemicelluloses are believed to be the primary foaming agents and to form a monomolecular layer around gas rumen bubbles that has its greatest stability at about pH 6.0. Salivary mucin is antifoaming, but saliva production is reduced with succulent forages. Bloat-producing pastures are more rapidly digested and may release a greater amount of small chloroplast particles that trap gas bubbles and prevent their coalescence. The immediate effect of feeding is probably to supply nutrients for a burst of microbial fermentation. However, the major factor that determines whether bloat will occur is the nature of the ruminal contents. Protein content and rates of digestion and ruminal passage reflect the forage's potential for causing bloat. Over a 24-hr period, the bloat-causing forage and unknown animal factors combine to maintain an increased concentration of small feed particles and enhance the susceptibility to bloat.

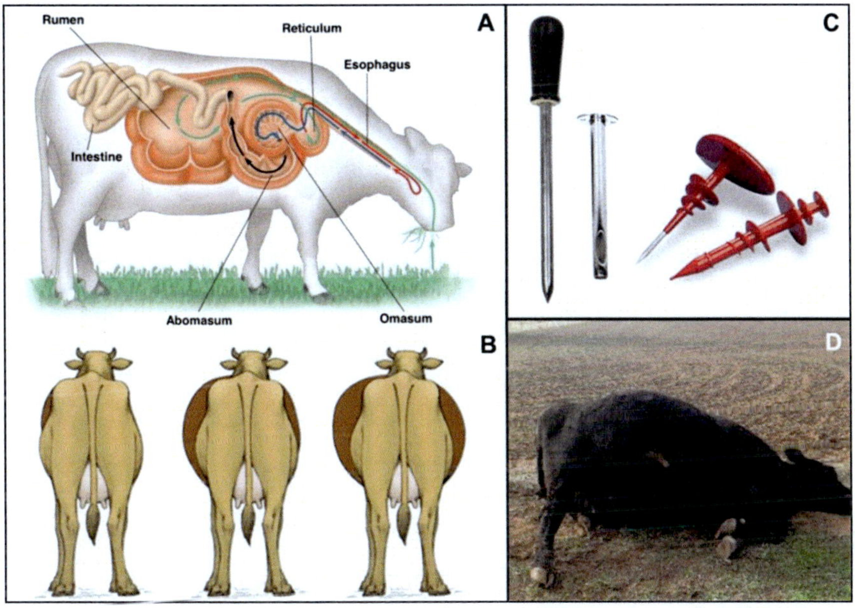

Figure 1. Pasture bloat: a big problem for cattle producers. (A) Ruminants have a four-compartment stomach: rumen, reticulum, omasum and abomasum. Adapted from Addison Wesley Longman Inc., 1999. (B) Visual signs of bloated cattle include mild (left), moderate (middle) or severe (right) distension of the left side of the animal. Illustration by Tom Milner, Out Here Magazine, Fall 2012. (C) In life-threatening cases of bloated livestock, a trocar (left) and cannula (righ) may be used for emergency relief. Different types of combined trocar and cannula can be used depending on each case. (D) Many cattle die by asphyxiation unless the bloat is not resolved by eructation, or by manual or surgical alleviation of the gas or fluid accumulated.

Pasture bloat usually occurs in animals grazing wheat legume or legume-dominant pastures, particularly alfalfa, ladino, and red or white clovers. Legume forages such as alfalfa and clover have a higher percentage of protein and are digested more quickly. Other legumes, such as sainfoin, crown vetch, milk vetch, fenugreek, and birdsfoot trefoil, are high in protein but do not cause bloat, probably because they contain condensed tannins (CTs), which precipitate protein and are digested more slowly than alfalfa or clover. Leguminous bloat is most common when cattle are placed on lush pastures, particularly those dominated by rapidly growing leguminous plants in the vegetative and early bud stages, but can also be seen when high-quality hay is fed.

Prevention of pasture bloat is difficult. Management practices used to reduce the risk of bloat include feeding hay (orchard grass) before turning cattle on pasture, maintaining grass dominance in the sward or using strip grazing to restrict intake, with movement of animals to a new strip in the afternoon, not the early morning. Hay must constitute at least one-third of the diet to effectively reduce risk of bloat. Feeding hay or strip grazing may be reliable when the pasture is only moderately dangerous, but these methods are less reliable when the pasture is in the pre-bloom stage and the bloat potential is high. Mature pastures are less likely to cause bloat than immature or rapidly growing pastures. The only satisfactory method available to prevent pasture bloating is continual administration of an antifoaming agent during the risk period. The most reliable method is drenching twice daily (at milking times) with an antifoaming agent. Spraying the agent onto the pasture is equally effective, provided the animals have access only to treated pasture. This method is ideal for strip grazing but not when grazing is uncontrolled. The antifoaming agent can be added to the feed or water or incorporated into feed blocks, but success with this method depends on adequate individual intake. The agent can be painted on the flanks of the animals, from which it is licked during the day, but animals that do not lick will be unprotected. Available antifoaming agents include oils and fats and synthetic nonionic surfactants. Oils and fats are given at 60-120 ml/head/day; doses up to 240 ml are indicated during dangerous periods. Poloxalene, a synthetic polymer, is a highly effective nonionic surfactant that can be given at 10-20 g/head/day and up to 40 g/head/day in high-risk situations. It is safe and economical to use and is administered daily through the susceptible period by adding to water, feed grain mixtures, or molasses. Pluronic agents facilitate the solubilization of water-insoluble factors that contribute to formation of stable foam. A pluronic detergent and a water-soluble mixture of alcohol ethoxylate and pluronic detergents also are effective. Ionophores (lasalocid, monensin) effectively prevent bloat, and a sustained-release capsule administered into the rumen and releasing 300 mg of monensin daily for a 100-day period protects against pasture bloat and improves milk production on bloat-prone pastures.

The ultimate aim in control of pasture bloat is development of a pasture that permits high production, while keeping incidence of bloat low. The use of pastures of clover and grasses in equal amounts comes closest to achieving this goal. Bloat potential varies between cultivars of alfalfa, and low-risk LIRD (Low Initial Rate of Digestion) cultivars are available commercially. The addition of legumes with high condensed tannins (or proanthocyanidins) to the pasture seeding mix (10% sainfoin) can reduce the risk of bloat where there is strip grazing, as can the feeding of sainfoin pellets (Lees, 1992). To prevent feedlot bloat, rations should contain ≥10-15% cut or chopped roughage mixed into the complete feed. Preferably, the roughage should be a cereal, grain straw, grass hay or equivalent. Grains should be rolled or cracked, not finely ground. Pelleted rations made from finely ground grain should be avoided. The addition of tallow (3%-5% of the total ration) may be successful occasionally, but it was not effective in controlled trials. The nonionic surfactants, such as poloxalene, have been ineffective in preventing feedlot bloat, but the ionophore lasalocid is effective in control.

1.2. Pasture Legumes and Condensed Tannins: The Alfalfa Case

About 153 species of forage legumes are currently cultivated worldwide (FAO 2015; www.feedipedia.org). Alfalfa (*Medicago sativa* L.) is one of the main forages grown in the world due to its high protein content. Alfalfa is a winter hardy, drought tolerant, N-fixing legume with a good longevity, high yield, high nutrient levels, high digestibility, unique structural to non-structural components ratio, high dry matter intake, and high animal productivity per hectare. However, its main limitation is its excessively rapid initial rate of protein degradation in the rumen, which results in pasture bloat and inefficient use of protein with consequent excessive excretions of nitrogen into the environment. Condensed tannins or proanthocyanidins are secondary plant metabolites that can bind with protein and thereby reduce the rate and extent of ruminal protein degradation. Unfortunately, these secondary metabolites do not accumulate in alfalfa (Barry and McNabb, 1999).

Oligomeric and polymeric proanthocyanidins are one of the end products of the flavonoid biosynthetic pathway (Dixon et al., 2005; 2012). These are present in the fruits bark, leaves and seeds of many plants where they provide protection against predation. PAs are very powerful antioxidants that remove harmful free oxygen radicals from cells. Studies indicate that antioxidant power of PAs is 20 times higher than that of vitamin C and 50 times higher than vitamin E. These findings, by no means, make PAs better antioxidants than vitamins C and E, but cannot replace them since vitamins play a crucial role in other functions and metabolic processes in the body. Unlike hydrolysable tannins, condensed tannins do not interfere with iron absorption. Food sources of PAs are grapes (seeds and skins), apples, unsweetened baking chocolate, red wines, blueberries, cranberries, bilberries, black currants, black chokeberries, hazelnuts, pecans or pistachios (Du et al., 2010). Over the past years PAs supplements have become popular in the form of oligomeric proanthocyanidin complexes (OPCs), grape seed extracts and maritime pine bark extracts. For instance, the introduction of PAs into alfalfa (*Medicago sativa*) forage is particularly important to ruminant livestock producers to combat pasture bloat. Currently, the livestock diet can be complemented with surfactants to break down the protein foams (Hall et al., 1994) or mixed with forage known to contain moderate levels of PAs (Li et al., 1996). Both are costly options.

The introduction of PAs into new alfalfa cultivars (*Medicago sativa*) can be particularly important to ruminant livestock producers. The alfalfa has high protein content (18-22% DW) but lacks PAs in their vegetative parts. The presence of PAs in the alfalfa plants could eliminate pasture bloat, improve the efficiency of conversion of plant protein into animal protein (ruminal protein bypass), reduce greenhouse gases, reduce gastrointestinal parasites and inhibit insect feeding (Waghorn, 1990; Niezen et al., 1995; Niezen et al., 1998; Aerts et al., 1999; Muir et al., 1999; McMahon et al., 2000). In the presence of PAs, excess dietary proteins as well as bacterial enzymes are complexed, and the level of protein degradation in the rumen is significantly reduced. This leads to an increase of protein bypass to the ruminant's gut and improved absorption of essential amino acids, resulting in increased milk and meat production. Moreover, decreased protein degradation in the rumen also decreases methane production and ammonium excretion in urine, which could contribute to significant reductions in emissions of the potent greenhouse gases methane and nitrous oxide from pastures. Classical breeding approaches have failed to introduce PAs into alfalfa foliage, and this problem is likely to require a biotechnological solution (Lees, 1992). Since PAs and anthocyanins (also lack in their aerial parts) share the same precursors, one approach to address this problem is to first transform plants with a transcription factor or factors which, when ectopically expressed, induce anthocyanidin production (Lloyd et al., 1992). Co-expression of one or more PA-specific biosynthetic enzymes such as anthocyanidin reductase (ANR), which converts cyanidin to the flavan-3-ol (-)-epicatechin, a likely building block of PAs (Dixon et al., 2005), may then lead to PA accumulation (Xie et al., 2006).

1.3. The Anthocyanin and Proanthocyanidin Pathways: Regulation and Intermediate Products

Three enzymes (leucoanthocyanidin reductase [LAR], anthocyanidin synthase [ANS], and anthocyanidin reductase [ANR]) function at branches between anthocyanin and PA biosynthesis. Proanthocyanidin biosynthesis (Figure 2A) utilizes two intermediates of the flavonoid biosynthetic pathway, leucoanthocyanidin for the formation of (+)-catechin and (+)-gallocatechin by means of LAR (Tanner et al., 2003), and anthocyanidin for the formation of (-)-epicatechin and (-)-epigallocatechin by means of ANR (Xie et al., 2003, 2004). Both of these intermediates are unstable and do not accumulate to measurable levels in plants (Peel et al., 2009). The ability to manipulate flavonoid biosynthesis in crop plants is gaining rapidly in importance as new economically uses are found in the areas of nutraceuticals and food and feed quality. Since PAs and anthocyanins share the same precursors, one approach to address this problem is to first transform plants with a transcription factor or factors which, when ectopically expressed, induce anthocyanidin production (Lloyd et al., 1992). Co-expression of one or more PA-specific biosynthetic enzymes such as anthocyanidin reductase (ANR), which converts cyanidin to the flavan-3-ol (-)-epicatechin, a likely building block of PAs (Dixon et al., 2005), may then lead to PA accumulation (Xie et al., 2006).

There have been many published reports of plant transcription factors (TFs) which regulate the flavonoid pathway in various tissues or in response to abiotic factors. The regulation of anthocyanin and PA biosynthesis is controlled by a ternary complex of transcription factors, called the MBW complex, consisting of MYB, basic helixloop-helix (bHLH), and WD40 transcription factors (Baudry et al., 2004; Albert et al., 2014). Among these, WD40 repeat transcription factors are required for both anthocyanin and PA biosynthesis. It is the MYBs, and to a lesser extent the bHLHs, that determine the specificity of gene activation for anthocyanin and PA biosynthesis. MYB and bHLH proteins generally act together to control various stages of the flavonoid pathway (Gong et al., 1999; Ramsay et al., 2003; Robbins et al., 2003; Deluc et al., 2006; Grotewold, 2006; Sweeney et al., 2006; Li et al., 2007; Park et al., 2007). These TFs regulate the expression of genes encoding the enzymes that catalyze each step of the biosynthetic pathway, such as chalcone synthase (CHS), chalcone isomerase (CHI), flavanone 3-hydroxylase (F3H), dihydroflavonol 4-reductase (DFR), anthocyanidin synthase (ANS), and UDP glucose-flavonoid 3-O-glucosyltransferase (UFGT).

The anthocyanin branch of flavonoid biosynthesis is activated in *Antirrhinum majus* flowers by two transcription factors, *AmRosea1* (encodes a MYB-related transcription factor) and *AmDelila (encodes a basic helix-loop-helix transcription factor),* (Schwinn et al., 2006). Tomato plants ectopically expressing *AmRosea1* and *AmDelila* in their fruit under the control of the ripening-induced E8 promoter have been shown to accumulate high levels of anthocyanin pigments (Butelli et al., 2008). Anthocyanin accumulation in *AmRos1/Am*Del transgenic fruit results from the activation of almost all of the genes encoding flavonoid biosynthetic enzymes. To date, functionality of both TFs has not been studied by stable transformation of other plant species (including legumes) so that it could be interesting to investigate their functionality in alfalfa to *de novo* activate anthocyanin biosynthesis.

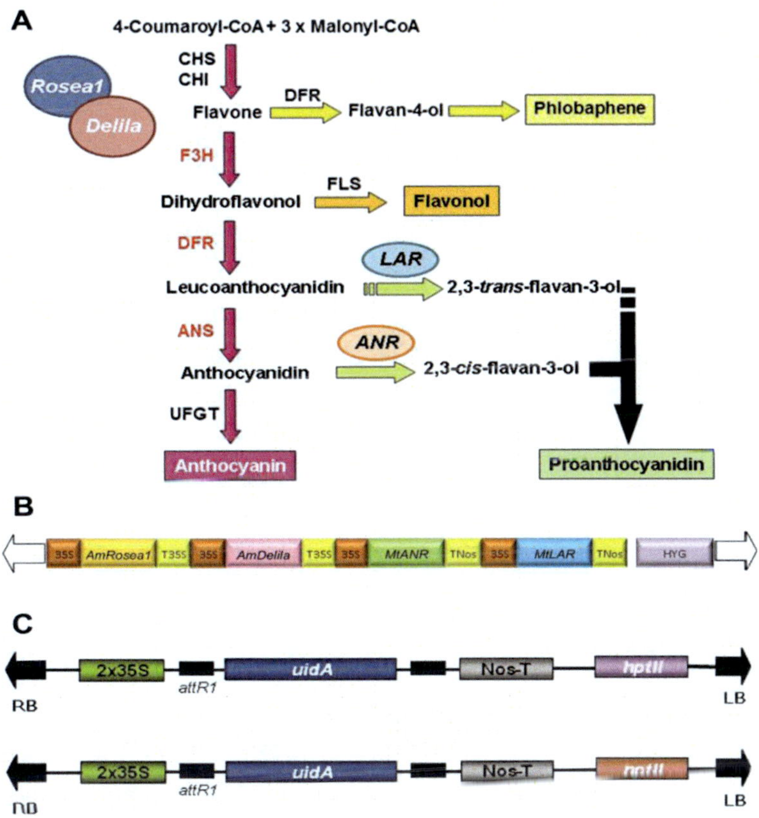

Figure 2. (A) Schematic representation of the biosynthetic pathway for anthocyanins and proanthocyanidins. Abbreviations: chalcone synthase (CHS); chalcone isomerase (CHI); dihydroflavonol reductase (DFR); flavanone 3-hydroxylase (F3H); flavonol synthase (FLS); leucoanthocyanidin reductase (LAR); anthocyanidin synthase (ANS); anthocyanidin reductase (ANR); uridine diphosphate glucose-flavonoid 3-*O*-glucosyl transferase (UFGT). (B) Multigenic construct *AmRosea1-AmDelila-MtANR-MtLAR* generated in the binary vector pDGD2. Abbreviations: CaMV35S promoter (35S); CaMV35S terminator (T35S); nopaline synthase terminator (Tnos); hygromicync resistance gene (*HYG*); *M. truncatula* anthocyanidin reductase gene (*MtANR*); *M. truncatula* leucoanthocyanidin reductase gene (*MtLAR*) (C) Constructs in the vector pMDC32 carrying the *uidA* (GUS-intrón) gene under the control of the CaMV35S (2X) constitutive promoter. Each construct carry the *hptII* and *nptII* selection marker genes conferring resistance to the antibiotics hygromycin and kanamycin respectively.

1.4. Genetic Regulation of the Anthocyanin and Proanthocyanidin Pathways in Legumes: What We Know

Transcriptional regulation of flavonoid biosynthesis is poorly understood in legumes. Major forage legumes like clovers, alfalfa and lupine do not contain appreciable amounts of PA's in the leaves (Li *et al.* 1996), where they only accumulate in glandular trichomes, being insufficient to prevent pasture bloat in ruminant livestock. These forage legumes show only very small natural variation, which makes it almost impossible to alter PA levels in these plants by conventional breeding methods (McMahon et al., 2000). It has been suggested that a PA concentration of about 20mg/g in the forage might be sufficient to prevent frothy bloat in cattle (Li et al., 1996). Attempts to alter PA levels in forage legumes either try to increase PAs to create 'bloat-safe' plants or to decrease PAs to improve nutritional value and palatability in potential crop species, however only few reports on the genetic modification of anthocyanins and condensed tannins in forage legumes have been published.

Different TFs have been proved to be functional in alfalfa and other legume species. The MYB transcription factor *LAP1* from *M. truncatula* induces constitutive production of anthocyanins when expressed in *M. sativa*, *M. truncatula* or white clover (Peel et al., 2009). Polymeric compounds probably derived from anthocyanidins accumulated in the two Medicago species, but these did not appear to be classical proanthocyanidins. Ectopic overexpression of the MYB *MtPAR*, another MYB transcription factor which redundantly regulates PA biosynthesis in the seed coat of *M. truncatula* led to massive accumulation of PAs in *M. truncatula* hairy roots as a result of increases in the expression of multiple genes including *CHS*, *F3H*, *ANS*, and *ANR*. However, transgenic alfalfa plants overexpressing *MtPAR* produced detectable but very low levels of PAs in the leaves (Verdier et al., 2011), indicating that *MtPAR* expression is sufficient to induce PAs accumulation in organs other than seed, but additional factors would be necessary to allow for high levels of PA accumulation. Similar approaches to induce PA levels in legume foliage have likewise been problematical. Recently, it has been reported that overexpression of *MtMYB5* or *MtMYB14* strongly induces PA accumulation in *M. truncatula* hairy roots, and both *myb5* and *myb14* mutants of *M. truncatula* exhibit darker seed coat colour than wild type plants. Moreover, they demonstrate that *MtMYB5* and *MtMYB14* physically interact and synergistically activate the promoters of anthocyanidin reductase and leucoanthocyanidin reductase, the key structural genes leading to PA biosynthesis, in the presence of *MtTT8* and *MtWD40-1* (Liu et al., 2014; Li et al., 2016).

1.5. Metabolic Engineering: Multigenic Constructs to Manipulate Complex Metabolic Pathways

The majority of valuable agronomic traits are polygenic, therefore the introduction of novel genetic pathways or the modification of the existing ones may require the simultaneous incorporation of several transgenes to the plant (Ruíz-López et al., 2012; Farre et al., 2013). In the last decades, technical hurdles limiting the number of genes transferred to plants have introduced a significant bottleneck to progress in plant biotechnology (Capell et al., 2004; Halpin et al., 2001, 2005; Dafny-Yelin et al., 2007). Early transformation methods for plants were developed with the implicit intention to introduce one or two genes (usually a primary transgene and a selectable or screenable marker) and have been optimized on that basis (Twyman et al., 2002). The genes are stacked in transgenic plants using iterative processes, such as successive rounds of crosses between different transgenic lines (Ma et al., 1995; Datta et al., 2002), or by means of the sequential transformation of transgenic plants with additional transgenes (Lapierre et al., 1999; Jobling et al., 2002; Qi et al., 2004). Both methods have two major drawbacks: the long and labor-intensive development process involving several breeding generations and the fact that the different transgenes are unlinked, leading to segregation in subsequent generations (Halpin et al., 2001). One successful strategy for the simultaneous transformation of different plant species with several genes at a time is the co-bombardment of several transgenes using biolistics. The different transgenes tend to integrate in the same locus and are jointly inherited (Naqvi et al. 2009). Multiple gene transfer has been mainly used for the analysis and modification of metabolic pathways, which can include large number of genes and complicated regulatory mechanisms (Naqvi et al., 2009; Jiang et al., 2013).

The alternative to unlinked transformation is the simultaneous transfer to the plant of several genetic units that were assembled into the same DNA string. This facilitates not only the inheritance of all the transgenes among generations but also the selection of a successful transformation event, because only one selection marker is required. Several methods have been implemented to facilitate the assembly of more than two genes in the same plasmid but most of them are limited in the number of parts to be assembled (Multisite Gateway) and others require *de novo* synthesis of parts in each assembly reaction to add overlapping regions limiting the standardization and reusability. An interesting alternative approach is the modular construction of genetic devices using standardized DNA parts. Modular design facilitates combinatorial engineering, as standard DNA parts can be easily exchanged improving the possibilities of the building process. A good example is the GoldenBraid2.0 (GB2.0) cloning system, a standardized DNA assembly platform developed to facilitate multigene engineering in plants by the simultaneous incorporation of several transgenes (https://gbcloning.upv.es/). GB2.0 allows the binary combination of multipartite assemblies using an extremely simple set of rules. This cloning system facilitates the reusability of DNA parts and assembled devices to efficiently build complex constructs and includes a collection of pre-assembled genetic devices that facilitate the generation of multigenic constructs (Sarrion-Perdigones et al., 2011; 2013; 2014). This multigenic cloning system has been extensively proved by transient expression (agroinfiltration) in *Nicotiana benthamiana* leaves and in the stable transformation of *N. benthamiana and N. tabacum* (Vafaee et al., 2014; Polturak et al., 2016).

2. MATERIAL AND METHODS

2.1. Plant Material

Alfalfa (*Medicago sativa*) Regen SY27, Regen 4D and Aragón plants, and *Medicago truncatula* 2HA plants were grown in a growth chamber with 16 h light/8 h dark photoperiod and temperature range of 22^{0}C day/18^{0}C night. *Arabidopsis thaliana* cv. Columbia (Col) plants were grown in growth chambers at 21^{0}C under long-day (16 h light) conditions, in a 1:1:1 mixture of sphagnum:perlite:vermiculite and illuminated by cool-white fluorescent lamps (150 μE m^{-2} s^{-1}). Arabidopsis plants were irrigated with water and, once a week, with a Hoagland N°.1 mineral solution supplemented with oligoelements (Hewitt 1966).

2.2. Bacterial strains and Growth Conditions

Escherichia coli DH5α was used for gene cloning. *Agrobacterium tumefaciens* strains AGL1 and LBA4404 were used for stable transformation experiments in alfalfa. Both strains were grown in LB medium under agitation (200 rpm) at 37^{0}C and 28^{0}C respectively. Ampicillin, kanamycin and spectinomycin were used for *E. coli* at 50 μg mL^{-1}. Rifampicin was also used for *A. tumefaciens* AGL1 and LBA4404 at 40 and 20 μg mL^{-1} respectively.

2.3. Protocol for Agrobacterium-Mediated Transformation of Alfalfa via Somatic Embryogenesis

2.3.1. Germination of Alfalfa Seeds

Put 20-30 alfalfa seeds into a 50 ml falcon tube, add 5 ml of concentrated sulphuric acid and left submerged during 5 minutes without shake. Remove the acid and add 45 ml of sterile distilled water. Upset the tube during two minutes and withdrew the water. Repeat the washes 3-4 times. Place a moistened filter paper (5-6 ml of water) in a Petri plate, put the seeds on damp paper and close the plate sealing it with plastic film. The plate is finally covered with aluminium foil and left at room temperature for 2-3 days. The *in vitro* germinated seedlings are then transferred to small pots with substrate (50% blond peat, 25% vermiculite nº 3 and 25% perlite) in the greenhouse. Cover the pots with a plastic film to keep the humidity conditions. After a week, make small holes in the film till plantlet acclimation. The developed plants will be used as explants source for the transformation assays.

2.3.2. Preparation of the A. tumefaciens AGL1 Inoculum

The protocol began 5 days before the co-cultivation phase growing in a plate of LB medium, with the corresponding antibiotic for selection, the *A. tumefaciens* strain AGL1 (preserved in stocks of glycerol at -80^{0}C) containing the desired construct. To get isolated colonies the plate is cultivated in stove at 28^{0}C during 72 hours. Two days before the co-cultivation, one isolated colony is inoculated in 3 ml of sterile LB medium with the corresponding antibiotic and later on cultivated overnight (o/n) at 28^{0}C in a shaker at 200 rpm. Using 1:300 v/v of this pre-inoculum, a flask of 250 ml containing sterile LB medium with the corresponding antibiotic, is inoculated with 25-40 ml of pre-inoculum and is left grow the bacteria through agitation of 200 rpm at 28^{0}C leaving it o/n to get to an OD_{600}: 1.0. The day of the infection, the bacteria culture is poured in a centrifuge tube of 250 ml and centrifuged at 6000 rpm for 10 minutes at 4^{0}C, after that the supernatant was carefully removed and discarded. Leave the pellet at 4^{0}C until use. Just before the infection the pellet was resuspended by gently shaking in liquid co-culture medium (CoM) containing 4.30 gL^{-1} MS basal salt mixture, 0.25 gL^{-1} casein hydrolysate, 1 mlL^{-1} Gamborg's B5 vitamins and supplemented with acetosyringone (100μM) to induce the T-DNA, until it came to an OD_{600}: 0.5.

2.3.3. Preparation of Leaf Explants and Co-Culture

Explants are collected from leaves of alfalfa plants grown under greenhouse conditions. At difference of other transformation protocols that use as starting explants leaves from *in vitro* propagated plants, our experience with *M. truncatula* showed us that the use of leaf explants from plants grown under greenhouse conditions is more convenient. These leaves are more developed, more strong and healthy to tolerate the infection process with the AGL1 strain of Agrobacterium. To prepare the leaf explants, in each experiment of transformation, we collected around 30 young leaflets that were completely stretched. From each trifoliolate leaf it can get 3 to 6 explants depending on the size of the leaf chosen. The leaves are washed with 10% bleach with a drop of Tween 20 during 15 min in sterile magenta jars (10 leaves by jar), and washed again with sterile water three times. The washed leaves were transferred to sterile Petri dishes to carry out its dissection. Initially, the three leaflets that compose a trifoliate leaf were separated by cutting their petioles with a scalpel and finally the leaflets were placed with the underside facing up on the plate. Then the leaves were cut with a sterile scalpel by removing the side edges to facilitate the entry of the bacteria in the plant tissue. Dissected leaves are incubated with the suspension of Agrobacterium containing the plasmid of interest in Petri dishes with co-culture (CoM) liquid medium containing 4.30 gL^{-1} MS basal salt mixture, 0.25 gL^{-1} casein hydrolysate, 1 mlL^{-1} Gamborg's B5 vitamins and supplemented with acetosyringone 100μM for 10 minutes. The CoM medium is sterilized by microfiltration and pH is adjusted to 5.7 with 1M NaOH.

Once finished that time, the leaf sections are blotted dry with sterile filter paper and placed in CoM solid medium without antibiotics and with the abaxial part of leaflets towards above. In this medium we replaced the kinetin as source of cytokinins as described other protocols by 6-benzylaminopurine (BAP). The inoculated explants are maintained during 5 days in darkness at 22^0C, rather than 24^0C as is described in other transformation protocols. By decreasing temperature, Agrobacterium growth slowly and damage of the explant is considerably reduced. Other protocols eliminated the excess of Agrobacterium in the explants by decontamination with water introducing the explants into a falcon with water and agitating the tube for 20 seconds. Passed this time, the explants are dried with sterile filter paper and placed it back at the same environmental conditions. This process it repeated every two days. We have recently changed this decontamination process after 5 days of co-culture, carrying out an incubation in liquid CoM supplemented with 250 mg/L^{-1} of ticarcillin (which inhibits the growth of Agrobacterium) o/n at 28^{0}C with agitation at 180-200 rpm. Using this approach we avoid the excessive handling and the corresponding damage of the explants, achieving the complete removal of Agrobacterium.

2.3.4. Callus Induction, Somatic Embryogenesis and Selection of Transformants

The leaf explants are transferred to solid callus induction medium (IcM) containing 4.30 gL^{-1} MS basal salt mixture, 0.25 gL^{-1} casein hydrolysate, 30 gL^{-1} sucrose, 1 mlL^{-1} Gamborg's B5 vitamins, 1 mlL^{-1} 10 mM 2,4-D, 0.5 mlL^{-1} 10mM BAP, 8 gL^{-1} Phytagel Sigma and supplemented with 250 mg/L of ticarcillin and with the selective agent according to the plasmid used (in our case 15 mgL^{-1} hygromycin or 50 mgL^{-1} kanamycin). Then the explants are incubated at 24^{0}C in soft light (30µE m^{-2} s^{-1}) and subcultured every two weeks. The IcM medium also contained a synthetic auxin (2, 4-dichlorophenoxyacetic acid; 2, 4-D) and a cytokinin (BAP) that allowed the dedifferentiation of plant cells and their proliferation. After a few days of cultivation a thickening of the court areas is produced, which derived in the formation of callus that finally is extended along the explant. In addition, with the transfer to light conditions, the callus took a yellow-green colour.

After four or five weeks of cultivation, calli began to form proembryogenic masses and then are transferred to embryo development medium (DeM) containing 4.30 gL^{-1} MS basal salt mixture, 0.25 gL^{-1} casein hydrolysate, 30 gL^{-1} sucrose, 1 mlL^{-1} Gamborg's B5 vitamins, 0.3 mlL^{-1} 10 mM 2,4-D, 7.5 gL^{-1} Phytagel Sigma and supplemented with 250 mg/L^{-1} of ticarcillin, the selective agent, and with 6 M proline, since it has been described that the addition of this aminoacid to the medium increases the production of somatic embryos starting from proembryogenic masses (Baeksed et al., 1997). The calli are cultivated with strong light (60-80 μE m^{-2} s^{-1}) at 24^{0}C in 16 h photoperiod conditions. Finally, the differentiation and subsequent development of globular embryos are produced.

2.3.5. Development and Germination of Embryos to Produce Plantlets

Four weeks after transfer to DeM medium, the formation of globular embryos can be observed. At this moment, developing embryos are transferred to embryo germination medium (GeM) containing 4.30 gL^{-1} MS basal salt mixture, 0.25 gL^{-1} casein hydrolysate, 1 mlL^{-1} Gamborg's B5 vitamins, 30 gL^{-1} sucrose, Phytagel Sigma 7.5 gL^{-1} and supplemented with 6 M proline, where the embryos continue their development. The calli are dispersed to put well in contact the developing embryos with the selective medium in order to avoid the appearance of escapes.

At this stage, we could separate embryos in cotyledonar stage to continue germinating by separate, or we could leave the embryos in this same medium until the root is developed 2-5 mm long and the first trifoliate leaf appears. At this time the alfalfa plantlets can be transferred to a plant development medium (DpM) containing 4.30 gL^{-1} MS basal salt mixture, 1 mlL^{-1} Gamborg's B5 vitamins, 30 gL^{-1} sucrose, Phytagel Sigma 7.5 gL^{-1} without proline and selective agent, in more high containers (Magenta type) to facilitate a proper root development and grow in height of the vegetative parts.

2.3.6. Acclimation of Transgenic Alfalfa Plants to Greenhouse Conditions

Rooted alfalfa plants are transferred from *in vitro* containers to small pots with substrate (50% blond peat, 25% vermiculite N° 3 and 25% perlite), when they had already formed 2-3 trifoliate leaves. The pots are covered with a plastic film to keep the 100% humidity conditions and were acclimated to environmental humidity progressively. Acclimation can be performed by opening small holes in the plastic wrap so that in a few days the plants close their stomata and adjust to the humidity of the greenhouse. The overall transformation protocol, since the co-culture of explants to the acclimation of plants lasts 20-25 weeks. It is suitable to keep a stock of plants *in vitro* by rooting of cuttings in a maintenance medium (SH10b; Chabaud *et al.* 2008). The aim of clonal propagation is to obtain identical plants to the original genotypes through vegetative multiplication (from axillary buds or apical meristems). Another advantage of this practice is the production of clones of a given transgenic plant of interest in order to have a great number of plants for further molecular or phenotypic analyses. Clones obtained by rooting in pots of axillary or apical buds from plants cultivated in the greenhouse can be also produced using a commercial mixture of rooting hormones (IBA / IAA).

2.4. Generation of the pDGB2-35S:*AmROS1*:T35S-35S:*AmDEL*:T35S-35S:*MtANR*:Tnos-35S:*MtLAR*:Tnos Multigenic Construct

Domestication of pDGB2 plasmid and generation of the multigenic construct was performed basically as earlier described (Engler et al., 2008; Sarrión-Perdigones et al., 2011). To generate the multigenic construct we have used two transcription factors: the R2-R3-MYB *AmROSEA1* and the bHLH *AmDELILA* from *Antirrhinum majus* in combination with the *anthocyanidin reductase* (*MtANR*) and the *leucoanthocyanidin reductase* (*MtLAR*) genes of *Medicago truncatula*, two enzymes that function at branches between anthocyanin and PA biosynthesis (Pang et al., 2007). The 35S *Cauliflower Mosaic Virus* (CaMV35S) constitutive promoter and terminator (T35S) were used for the *AmDel* and *AmRos1* genes. In the *MtANR* and *MtLAR* genes, the *nopalin synthase* terminator (Tnos) was used instead of the T35S (Figure 2B). The coding sequences of *MtANR* and *MtLAR* were adapted to the GB2.0 grammar as described earlier (Sarrion-Perdigones et al., 2013). *MtANR* was isolated by RT-PCR from RNA of *M. truncatula* seeds 10 days after pollination (reference number, NCBI: XM_013601695.1) using appropriate primers (Table 1). *MtLAR* was originally cloned from the legume *Desmodium uncinatum* (Tanner et al.,

2003) and we amplified the *M. truncatula* ortholog. Constructs were confirmed by plasmid restriction analysis and by sequencing.

2.5. Stable Genetic Transformation of *Arabidopsis thaliana*

Arabidopsis cv. Columbia plants were transformed according to standard procedures (Bechtold and Pelletier, 1998). The binary vector *pDGB2-35S:AmROS1:T35S-35S:AmDEL:T35S-35S:MtANR:Tnos-35S:MtLAR:Tnos* was electroporated into *A. tumefaciens* strain GV3101::pMP90(RK). Resulting seeds were harvested and surface-sterilized in 96% EtOH and alkylbenzyldimethylammonium chloride (0.01 gL^{-1}) for 10 min at 65^0C, soaked for 10 min in 7% NaClO, 1% SDS and finally, washed five times in sterile distilled water. Seeds were plated on the selection medium containing kanamycin (50 mgL^{-1}) and timentine (300 mgL^{-1}) to identify transgenic plants. Stratification of seeds was conducted during two days at 4^0C and after the seeds were incubated in a controlled environment growth chamber at 22^0C under a 16 h light/8 h dark photoperiod. For each construct, resistant kanamycin lines were used for phenotypic and molecular characterization.

2.6. DNA and RNA Extraction and Purification

Immature seeds were collected from *M. truncatula* pods at developmental stage of 10 days after anthesis using a fluorescence stereomicroscope (Leica, MZ16F). The pods were collected and kept on ice for not longer than 1 h and finally stored at -80^0C. Immature seeds were pulverized and suspended in ice-cold RNA lysis solution of a Plant RNA Purification Reagent (Invitrogen). Later on, purified RNA was obtained as described in the Turbo DNA-free instructions (Ambion) after remove traces of contaminating DNA. The cDNA samples were synthesized from 1 µg of total RNA using the Primescript™ RT Reagent Kit (Perfect Real

Time, Takara) and random primers in 20 µL total volume reaction. This cDNA was used to isolate the *MtANR* and *MtLAR* genes.

Total RNA was purified from tissue samples using the E.Z.N.A. Plant RNA Kit, following the manufacturer's recommendations (Omega). The amount of starting sample was 100 mg RNA quantified at 260 nm absorbance, whereas its purity was assessed based on absorbance ratios at 260/280 nm. The integrity of purified RNA was confirmed by denaturing agarose gel electrophoresis and ethidium bromide staining. Then, 1 µg of total RNA were reverse transcribed with a Primescript™ RT Reagent Kit (Perfect Real Time, Takara).

Genomic DNA was isolated from young leaves (100 mg) using the E.Z.N.A. Plant DNA Kit (Omega) according to the manufacturer's instructions and checked by standard agarose electrophoresis.

2.7. Semiquantitative RT-PCR

All semiquantitative RT-PCR reactions were performed and analyzed on a Thermo Scientific Arktik Thermal Cycler (Thermo Fisher Scientific). PCRs were performed taking 2 µL aliquots of cDNA solution in a 25 µL total volume reaction containing 0.2 mM of each dNTP, 3 mM $MgCl_2$, 1x reaction buffer,1 unit of Taq DNA Polymerase (Biotools), and 0.8 mM of the appropriate pair of primers (Table 1).

PCR conditions for amplification of *AmRosea1, AmDelila*, *MtANR* and *MtLAR* consisted of initial denaturation at 94^{0}C for 5 min, followed by 30 cycles of 94^{0}C/30 s, 55^{0}C/30 s, and 72^{0}C/60 s, and finally 7 min extension at 72^{0}C. For the *UBIQUITINE* gene 25 cycles were achieved. Twenty microliter of PCR products were separated on 1% ethidium bromide-stained agarose gels. PCR fragments were visualized under UV using a Syngene GBOX (Syngene) and captured with the GeneSnap program (Syngene). The analyses were carried out in duplicate using biologically independent material with similar results. The possibility of genomic DNA contamination in the RT-PCR assays was controlled with the No Reverse Transcriptase control (No RT control).

Table 1. Primers used in this work

Gene	Forward 5´-3´	Reverse 5´-3
AmDELILA	GAGGAAGCAACTTGCTATTGCTG	CATTGATGCCTTCTGCCAAATTCG
AmROSEA1	GGTAACAAATGGTCGCTGATTGC	CCATCCTCCGTTGTTTCTAGC
MtANR	GCATGTGTGATAGGTGGAC	GACCAGTCCCTTCGAGTTCG
MtLAR	GGTCGAGTCCTAATTGTTGG	CGTGACCAAATTCTGAAGGC
qRT-AmDELILA	TACGGACAAGATTGGAGCCG	GGAATCTTTCAATCGACCTCTG
qRT-AmROSEA1	GACTAATATCGTAAGACCCCGAGC	GTAACGTGCAATCCGGTGAAGG
qRT-MtANR	CAACTTCTGGTCGATACATTTGC	CGGGAACACTGGTATTGTGAG
qRT-MtLAR	CCACCGTTGGATCAATTGCAC	GTACGCTTTGACATTGCCATG
AtUBIQ	GATCTTTGCCGGAAAACAATTGGA GGATGGT	CGACTTGTCATTAGAAAGAAAGA GATAACAG

3. Results and Discussion

3.1. Biotechnological Approaches to *de novo* Induce the Anthocyanin and Proanthocyanidin Pathways in Alfalfa

We are currently developing biotechnological tools to *de novo* induce the anthocyanin and proanthocyanidin pathways in alfalfa. Our goal is to develop an efficient transformation method to produce transgenic alfalfa plants and to generate a multigenic construct to introduce into the plant the necessary genes to simultaneously activate both pathways in this forage legume.

3.2. Developing an Efficient Agrobacterium-Mediated Transformation Method to Produce Alfalfa Transgenic Plants

We have developed an efficient and reproducible *Agrobacterium*-mediated transformation protocol by induced somatic embryogenesis in leaf explants from the alfalfa Regen-SY27 genotype (Bingham 1991) propagated *in vitro* and acclimated in the greenhouse. Recently we have incorporated to our transformation assays the RSY4D genotype (The

Samuel Roberts Noble Foundation, USA) and the genotype Aragon, the most cultivated in Spain. This protocol is based in our previous experience producing transgenic *Medicago truncatula* plants (Trinh et al. 2001; Cosson et al. 2006, 2015; Chabaud 2008) and the protocols described by Samac et al. (2004, 2006) to genetically transform *Medicago sativa* plants. The protocol has been described in detail in the Material and Methods section. In figure 3A-I, the different stages of the protocol to produce transgenic alfalfa plants via somatic embryogenesis are shown.

Most of the alfalfa transformation methods published previously used the *Agrobacterium tumefaciens* LBA4404 strain as bacterial vector to introduce foreign genes into the plant, and the *nptII* (*neomycin phosphotransferase*) gene, which confers resistance to the antibiotic kanamycin, as selectable marker. We have recently incorporated to the transformation assays the AGL1 strain of Agrobacterium (Lazo 1991), which is more infective than the LBA4404. Most of these publications indicated that selection with 50 mgL^{-1} of kanamycin produces scapes, resulting in false transformants or chimaeric plants in which some of the introduced transgenes are absent. In our previous experience producing *Medicago truncatula* transgenic plants, we corroborated this fact and also that the use of high concentrations of this antibiotic were, in most cases, lethal for the regenerating plants. For this reason, we have been trying the *hptII* (*hygromycin phosphotransferase II*) gene as selectable marker in our constructs because it is less toxic for the plant and can be used at low concentrations (15 mgL^{-1}) than kanamycin (Cosson et al. 2007; Chabaud et al. 2008). Different transformation assays were performed using as starting material leaf explants from the Regen-SY27 genotype and combinations of both Agrobacterium strains (LBA4404 and AGL1) carrying two types of gene constructs, one with the *uidA* (GUS) reporter gene and the *nptII* gene as selectable marker, and other with the *uidA* gene and the *hptII* gene as selectable marker (Figure 3J-O). Our results indicated that the AGL1 strain in combination with the *hptII* gene (hygromycin resistance, 15 mgL^{-1}) was more efficient during the different regeneration phases (callogenesis, proembryogenesis and embryogenesis), in comparison with the LBA4404 strain with the *nptII* gene (kanamycin resistance, 50 mgL^{-1}), (Figure 4A).

Figure 3. Different stages of the *Agrobacterium*-mediated protocol for the genetic transformation of alfalfa via somatic embryogenesis. (A) Proembryogenic callus. (B) Embryogenic callus. (C) Embryogenic callus in germination medium. (D) Close view of an embryogenic callus showing different stages of embryo development (globular, heart, torpedo and cotyledonar). (E) Group of embryos in cotyledonar stage. (F) Isolated embryo before germination showing the radicle. (G) Alfalfa plantlets after embryo germination. (H) Acclimation of rooted alfalfa plants in the greenhouse (I) Transgenic alfalfa plant in a pot after acclimation. (J) Non-transformed Regen SY-27embryogenic callus (control). (K) Transgenic proembryogenic callus showing GUS expression (blue) and hygromycin resistance. (L) Proembryogenic callus showing a transgenic area (blue) in comparison with non-transgenic one (green) in a medium containing kanamycin as selective antibiotic. (M) Embryogenic callus with transgenic embryos (blue). (N) Germination of transgenic embryos (blue). (O) Transgenic embryo (blue) in cotyledonar stage.

The evaluation of the effectiveness of this *Agrobacterium*-mediated transformation protocol has been done in the context of a research project carried out over four years in our laboratory transforming three alfalfa varieties (Regen-SY27, RSY4D and the commercial genotype Aragon), with different gene constructs introduced via *A. tumefaciens* AGL1 and using the *hptII* gene (hygromycin resistance) as selective marker. Data corresponding to the three stages of embryo regeneration by somatic embryogenesis (callogenesis, proembryogenesis and embryogenesis), provided by different transformation assays have been used for the estimation of the protocol efficiency. The formation of calli (callogenesis) occurred between the 91-100% of the inoculated explants and in proembriogenesis we observed efficiencies ranging the 60-80% for the three genotypes (Figure 4B). The formation of viable embryogenic calli resulted in an efficiency of the 50% both for the Regen-SY27 and the RSY4D genotypes. This percentage refers to the formation of embryos in the cotyledonar (mature) phase, who continued their development until the production of complete rooted plants. In the case of the commercial genotype Aragon, the percentage of embryos that produced complete plants was 13% (Figure 4B). The transformation efficiency measured as the production of complete transgenic plants from an initial explant is difficult to define due to the variability among the different trials of transformation. This variability is possibly determined by differences in the production and maturation of embryos depending on the physiological state of the starting material and differences that sometimes showed the mother plants of a same genotype in terms of their capacity for regeneration. Thus, we can conclude that the 50% of the explants inoculated had the ability to regenerate mature embryos via somatic embryogenesis, and the 80-100% of plants obtained from these embryos (Figure 4C-D) contained the T-DNA with the complete set of the introduced transgenes.

In order to track the transformation protocol and evaluate its efficiency, we performed two constructs in the vector pMDC32 containing the *uidA* (GUS-intron) gene under the control of a double CaMV35S constitutive promoter. These constructs also carried the *hptII* gene and the *nptII* gene conferring resistance to hygromycin and kanamycin respectively (Figure 2C).

In Figure 3J-O we can observe some GUS stained tissues corresponding to different stages of the regeneration of transgenic embryos from proembryogenic calli. We note that both constructs with their respective selection agents resulted in transgenic embryogenic and proembryogenic masses. As we previously observed in *M. truncatula*, in the alfalfa calli developed in kanamycin containing media, a high number of false transformants are produced, reaching a ratio of 1:5 with respect to the hygromycin-resistant calli.

3.3. Generation and Validation of the Multigenic Construct *AmRosea1:AmDelila:MtANR:MtLAR* by Stable Genetic Transformation of *Arabidopsis thaliana*

Using the two transcription factors described by Butelli et al. (2008) to overproduce anthocyanins in tomato, *AmRosea1* (MYB) and *AmDelila* (bHLH), we have generated the pDGBΩ1_Hph_35S-*AmRosea1*-T35S::35S-*AmDelila*-T35S::35S-*MtANR*-Tnos::35S-*MtLAR*-Tnos multigenic construct in the binary vector pDGB2, based on the pCAMBIA series (Roberts et al., 1997), to activate the anthocyanin pathway in combination with two *M. truncatula* (*MtLAR* and *MtANR*) genes to produce the enzymes that will allow to proanthocyanidins (Figure 5).

Once the multigenic construct was generated, we achieved its experimental validation through stable genetic transformation of the model plant *Arabidopsis thaliana*. As a result of the *in planta* transformation of *A. thaliana* plants, we selected 10 independent transgenic plants carrying the construct *AmRosea1:AmDelila:MtANR:MtLAR* which were able to grow under hygromycin selection. When compared with *A. thaliana* wild type control plants (Figure 6A), seven of them showed purple pigmentation in the petiole of leaves and stems (Figure 6B), two presented intense purple

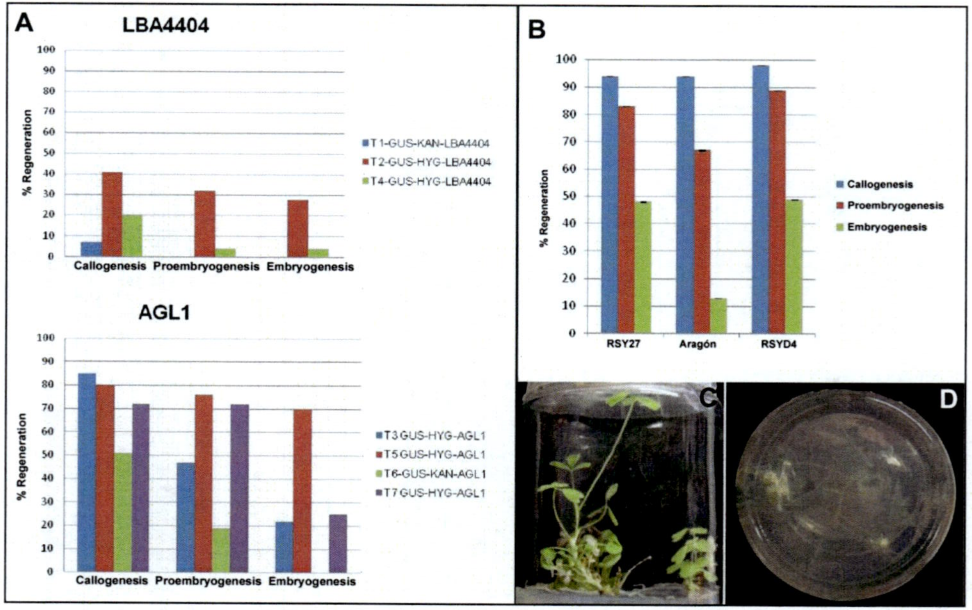

Figure 4. (A) Regeneration efficiency after combination of two Agrobacterium strains (LBA4404 and AGL1) with two selectable gene markers (*nptII*: kanamycin resistance and *hptII*; hygromycin resistance). Values represent the percentage of calli regenerated in the different phases of somatic embryogenesis: callogenesis, proembryogenesis and embryogenesis. The combination of the AGL1 strain with the resistance to hygromycin produces the highest regeneration rate. (B) Results of different transformation events to evaluate the efficiency of the protocol in three alfalfa genotypes: RSY 27, Aragón and RSY 4D, using the AGL1 strain of Agrobacterium and the *hptII* gene (hygromycin resistance). Values represent the percentage of regeneration in the different stages of somatic embryogenesis (callogenesis, proembryogenesis and embryogenesis) in the three alfalfa genotypes. The combination of the AGL1 strain with the resistance to hygromycin in the RSY 4D genotype produces the high regeneration rate, followed by the RSY 27 and Aragón genotypes. (C) *In vitro* rooted alfalfa plantlets after embryo germination. (D) Detailed view of roots at the base of transgenic alfalfa plantlets.

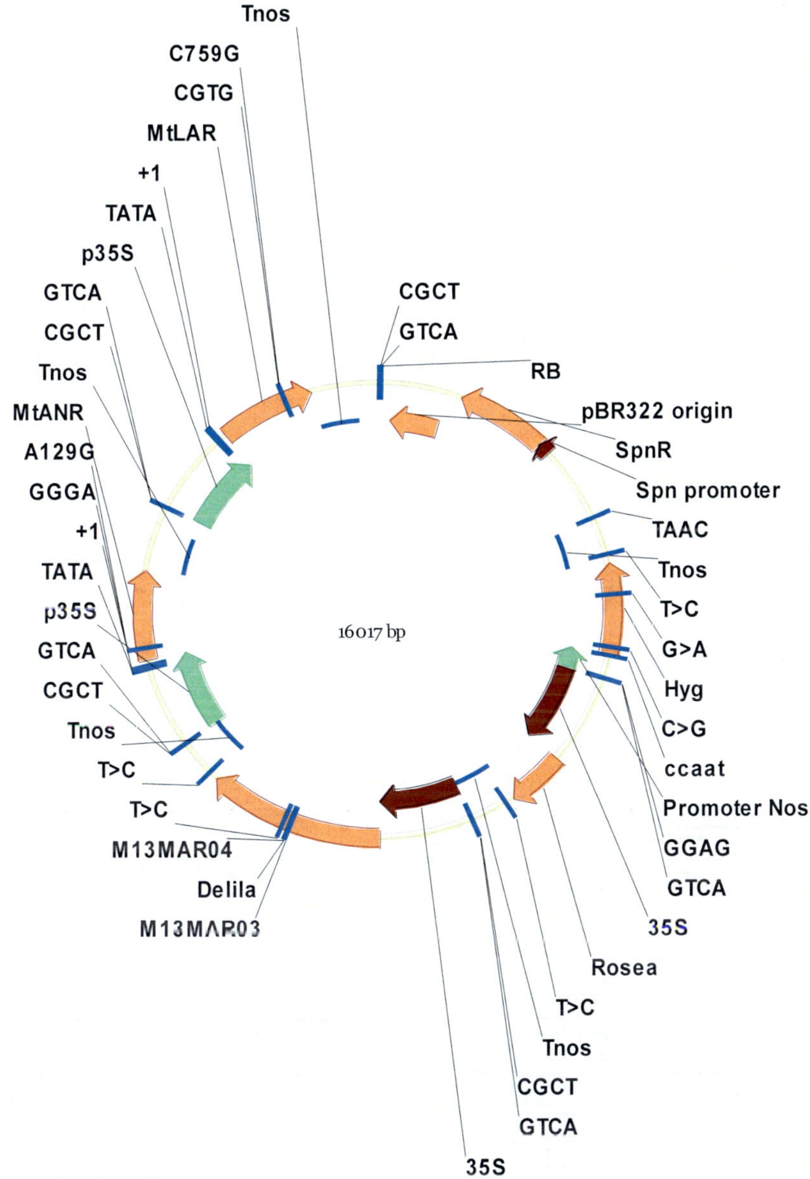

Figure 5. Schematic representation of the pDGBΩ1_Hph_35S-*AmRosea1*-T35S::35S-*AmDelila*-T35S::35S-*MtANR*-Tnos::35S-*MtLAR*-Tnos multigenic construct.

pigmentation in the different plant tissues (Figure 6C), indicating the *de novo* synthesis of anthocyanins in all the vegetative organs, and one plant (#8) showed no changes with respect to control plants (Figure 6D).

To study the integration capacity in the Arabidopsis genome of the four transgenes present in the multigenic construct, we performed a PCR from genomic DNA isolated from rosette leaves of the selected hygromycin-resistant plants using specific oligonucleotides for these genes (Table 1). Our results showed the presence of the four transgenes in 9 of the 10 plants analyzed. Only in one of the selected plants (#8), the four transgenes were absent (Figure 6E), corroborating that this plant was an escape because only the *hptII* gene was incorporated to its genome. This plant showed no changes of coloration in their vegetative organs, being its phenotype identical to the control plants (Figure 6D).

On the other hand, the expression level of the four transgenes were evaluated by semi-quantitative RT-PCR using RNA of cauline leaves from four transgenic plants (#2, 4, 10 and 11) and as a constitutive gene the endogenous *UBIQUITIN* gene (*AtUBIQ*). Our results showed that the expression level of the four transgenes evaluated separately, do not significantly vary among the different transgenic lines tested. However, we detected differences among the expression levels of each transgene in the transgenic plants. For instance, in plant #4, *AmRosea1* and *MtLAR* showed higher levels than *MtANR* and *AmDelila*. In the wild type control plants (WT) expression of the four transgenes was not detected (Figure 6F-G). We can conclude that the overexpression of the *Am Rosea1* and *AmDelila* TFs in the *A. thaliana* transgenic plants is related with the observed purple pigmentation in the plants, which is due to the activation of the anthocyanin pathway. Additional biochemical analyses would be necessary to determine if the expression of the *MtANR* and *MtLAR* genes in these plants is able to produce PAs using the two intermediates of the flavonoid biosynthetic pathway, leucoanthocyanidin for the formation of (+)-catechin and (+)-gallocatechin by means of *MtLAR*, and anthocyanidin for the formation of (-)-epicatechin and (-)-epigallocatechin by means of *MtANR*.

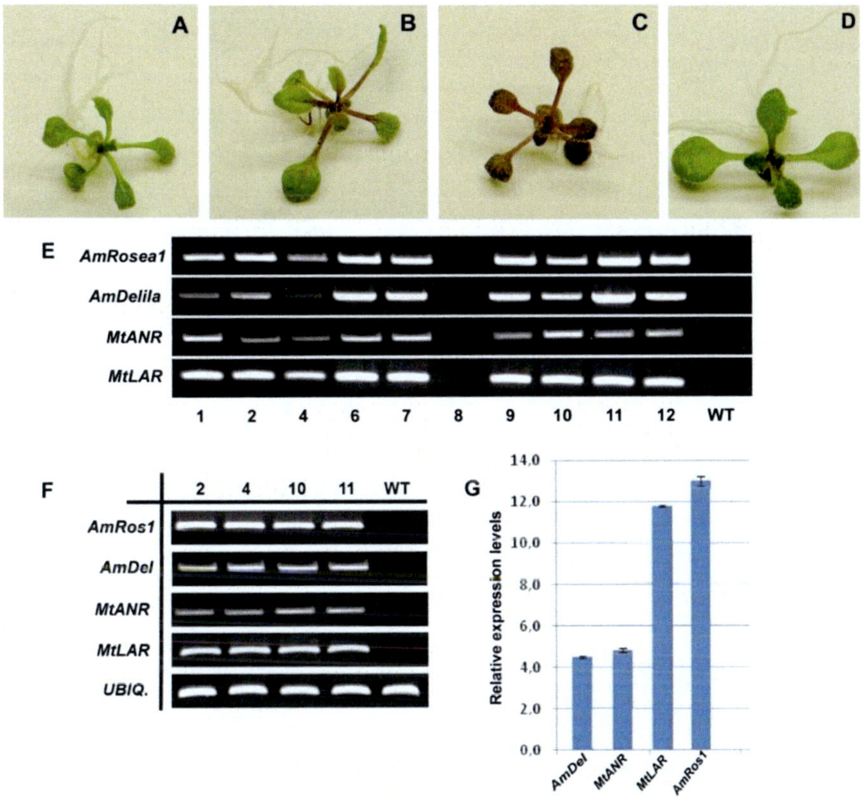

Figure 6. Phenotypes of the *AmRosea1:AmDelila:MtANR:MtLAR* transgenic *Arabidopsis thaliana* plants. (A) WT non-transgenic control plant. (B) Transgenic plant #2 showing purple pigmentation in the leaf petioles and stem. (C) Transgenic plant #10 constitutively showing intense purple pigmentation in all plant tissues. (D) Transgenic plant #8 showing no pigmentation. (E) Transgene integration in *A. thaliana.* Detection by PCR of the presence of the *AmRosea1, AmDelila, MtANR* and *MtLAR* transgenes in *A. thaliana* transgenic plants. Nine (plants #1, 2, 4, 6, 7, 9, 10, 11 and 12) of 10 plants checked contained the complete set of transgenes. The plant #8 was a scape and did not contain any of the four transgenes. (F) Semi-quantitative RT-PCR using RNA from cauline leaves of four *A. thaliana* transgenic plants (#2, 4, 10 and 11). (G) Expression levels of the *AmDelila*, *AmRosea1*, *MtANR* and *MtLAR* genes in transformed leaves of plant #4. Semi-quantitative RT-PCR results were obtained after 30 amplification cycles for all genes and 25 cycles for the *UBIQUITINE* gene.

Conclusions and Perspectives

New and improved alfalfa (*Medicago sativa*) cultivars have been obtained in the last decades by conventional breeding programmes. Biotechnology, genetic and metabolic engineering have been considerably developed and currently can provide molecular tools to the improvement of forage legumes. The alfalfa plant has high protein content but lacks condensed tannins or proanthocyanidins (PAs) in their vegetative parts. The presence of PAs in the alfalfa plants could prevent pasture bloat, improve the efficiency of conversion of plant protein into animal protein (ruminal protein bypass), reduce greenhouse gases, reduce gastrointestinal parasites and inhibit insect feeding. Pasture bloat usually occurs in animals grazing wheat legume or legume-dominant pastures, particularly alfalfa, ladino, and red or white clovers. Prevention of pasture bloat is difficult and death rates as high as 20% are recorded in cattle grazing bloat-prone pasture, and in pastoral areas, the annual mortality rate from bloat in dairy cows may approach 1%. There is also economic loss from depressed milk production in nonfatal cases and from suboptimal use of bloat-prone pastures. The introduction by metabolic engineering of the genes involved in the anthocyanin and PA biosynthesis into these new cultivars would be particularly important to ruminant livestock producers. Preliminary results obtained by our group to *de novo* produce PAs in the alfalfa forage are promising since we have developed two biotechnological tools essentials to achieve our objective: an efficient genetic transformation method to introduce foreign genes into different alfalfa cultivars and a multigenic construct able to introduce the four transgenes in the plant genome to *de novo* produce anthocyanins (the PA's precursors) in transgenic Arabidopsis plants. Stable transformation assays of different alfalfa genotypes with our multigenic construct are currently in progress to produce transgenic alfalfa plants with improved nutritional value to avoid pasture bloat in the ruminant livestock. After the problem of mad cow disease few years ago, the production of more safe forages to feed livestock is a main objective of most research projects worldwide. The production of forage legumes with high protein content and discrete levels of condensed tannins to reduce the occurrence of bloat and, at the same time, promote increased dietary protein nitrogen utilization avoiding problems of palatability is a challenge that can be achieved by metabolic engineering in a near future.

ACKNOWLEDGMENTS

This work was supported by grants BIO2012-39849-C02-01 and BIO2016-75485-R from the Spanish Ministry of Economy and Competitiveness (MINECO). We wish to thank Drs. Alejandro Sarrión-Perdigones and Diego Orzáez (IBMCP, Spain) for valuable comments on the GB2.0 technology. Technical assistance of Rafael Martínez-Pardo in the greenhouse is gratefully acknowledged.

REFERENCES

Aerts, R.J., Barry, T.N. and McNabb, W.C. 1999. Polyphenols and agriculture: beneficial effects of proanthocyanidins in forages. *Agric Ecosyst Environ* 75: 1-12.

Albert, N.W., Davies, K.M., Lewis, D.H., Zhang, H., Montefiori, M., Brendolise, C., Boase, M.R., Ngo, H., Jameson, P.E. and Schwinn, K.E. 2014. A conserved network of transcriptional activators and repressors regulates anthocyanin pigmentation in eudicots. *Plant Cell* 26: 962–980.

Barry, T.N. and McNabb, W.C. 1999. The implications of condensed tannins on the nutritive value of temperate forages fed to ruminants. *Br. J. Nutr.* 81: 263-272.

Baudry, A., Heim, M.A., Dubreucq, B., Caboche, M., Weisshaar, B. and Lepiniec L. 2004. TT2, TT8, and TTG1 synergistically specify the expression of *BANYULS* and proanthocyanidin biosynthesis in *Arabidopsis thaliana. Plant J.* 39: 366–380.

Bechtold, N. and Pelletier, G. 1998. *In planta* Agrobacterium-mediated transformation of adult *Arabidopsis thaliana* plants by vacuum infiltration. *Methods Mol Biol*, 82: 259-266.

Bingham, G.J. 1991. Registration of Alfalfa Hybrid Regen-SY Germplasm for Tissue Culture and Transformation Research. *Crop Sciencie* 31: 1098.

Butelli, E., Titta, L., Giorgio, M., Mock, H.-P., Matros, A., Peterek, S., Schijlen, E.G.W.M., Hall. R.D., Bovy, A.G., Luo, J. and Martin, C. 2008. Enrichment of tomato fruit with health-promoting anthocyanins by expression of select TFs. *Nat Biotechnol.* 26:1301-1308.

Capel,l T. and Christou, P. 2004. Progress in plant metabolic engineering. *Curr. Opin. Biotechnol.* 15: 148–154.

Chabaud, M., Boisson-Dernier, A., E.-P., de Carvalho-Niebel, F. and Barker, D.G. 2008. Agrobacterium-mediated transformation of *Medicago truncatula*. In *Handbook of new Technologies for genetic Improvement of legumes.* pp. 45-67.

Cosson, V., Durand, P., d'Erfurth, I., Kondorosi, A. and Ratet, P. 2006. *Medicago truncatula* transformation using leaf explants. *Methods Mol Biol*, 343: 115-127.

Cosson, V., Eschstruth, A. and Ratet, P. 2015. *Medicago truncatula* transformation using leaf explants. *Methods Mol Biol*, 1223: 43-56.

Dafny-Yelin, M. and Tzfira, T. 2007. Delivery of multiple transgenes to plant cells. *Plant Physiol.* 145: 1118-1128.

Datta, K., Baisakh, N., Maung Thet, K., Tu, J. and Datta, S.K. 2002. Pyramiding transgenes for multiple resistance in rice against bacterial blight, yellow stem borer and sheath blight. *Theor. Appl. Genet.* 106: 1-8.

Deluc, L., Barrieu, F., Marchive, C., Lauvergeat, V., Decendit, A., Richard, T., Carde, J.-P., Merillon, J.-M. and Hamdi, S. 2006. Characterization of a grapevine R2R3-MYB transcription factor that regulates the phenylpropanoid pathway. *Plant Physiol.* 140: 499- 511.

Dixon, R.A., Xie, D.Y. and Sharma, S.B. 2005. Proanthocyanidins: a final frontier in flavonoid research?. *New Phytol* 165: 9-28.

Dixon, R.A., Liu, Ch. and Jun, J.H. 2012. Metabolic engineering of anthocyanins and condensed tannins in plants. *Current Opinion in Biotechnology* 24: 1-7.

Du, H., Huang, Y. and Tang, Y. 2010. Genetic and metabolic engineering of isoflavonoid biosynthesis. *Appl Microbiol Biotechnol*, 86: 1293-1312.

Engler, C., Kandzia, R. and Marillonnet, S. 2008. A one pot, one step, precision cloning method with high throughput capability. *PLoS One* 3: e3647.

Farre, G., Maiam Rivera, S., Alves, R., Vilaprinyo, E., Sorribas, A., Canela, R., Naqvi, S., Sandmann, G., Capell, T., Zhu, C., Christou, P. 2013. Targeted transcriptomic and metabolic profiling reveals temporal bottlenecks in the maize carotenoid pathway that may be addressed by multigene engineering. *Plant J.* 75(3): 441-55.

Gong, Z.Z., Yamagishi, E., Yamazaki, M. and Saito, K. 1999. A constitutively expressed Myc-like gene involved in anthocyanin biosynthesis from *Perilla frutescens*: molecular characterization, heterologous expression in transgenic plants and transactivation in yeast cells. *Plant Mol. Biol.* 41: 33–44.

Grotewold, E. 2006. The genetics and biochemistry of floral pigments. *Annu. Rev. Plant Biol.* 57: 761-780.

Hall, J., Walker, I. and Majak, W. 1994. Evaluation of two supplements for the prevention of alfalfa bloat. *Can. Vet. J.* 35: 702-705.

Halpin, C., Barakate, A., Askari, B., Abbott, J. and Ryan, M.D. 2001. Enabling technologies for manipulating multiple genes on complex pathways. *Plant Mol. Biol.* 47: 295–310.

Halpin, C. 2005. Gene stacking in transgenic plants-the challenge for 21st century plant biotechnology. *Plant Biotechnol. J.* 3: 141–155.

Hewitt, E.J. 1966. Sand and water culture methods used in the study of plant nutrition. Farnham Royal, UK: Commonwealth Agricultural Bureau.

Jiang, L., Yu, X., Qi, X., Yu, Q., Deng, S., Bai, B., Li, N., Zhang, A., Zhu, C., Liu, B. and Pang, J. 2013. Multigene engineering of starch biosynthesis in maize endosperm increases the total starch content and the proportion of amylose. *Transgenic Res.* 22(6): 1133-42.

Jobling, S.A., Westcott, R.J., Tayal, A., Jeffcoat, R. and Schwall, G.P. 2002. Production of a freeze–thaw-stable potato starch by antisense inhibition of three starch synthase genes. *Nat. Biotechnol.* 20: 295-299.

Jonker, A. and Yu, P. 2016. The Role of Proanthocyanidins Complex in Structure and Nutrition Interaction in Alfalfa Forage. *Int. J. Mol. Sci.* 17: 793-810.

Lapierre, C., Pollet, B., Petit-Conil, M., Toval, G., Romero, J., Pilate, G., Leplé, J.C., Boerjan, W., Ferret, V., De Nadai, V. and Jouanin, L. 1999. Structural alterations of lignins in transgenic poplars with depressed cinnamyl alcohol dehydrogenase or caffeic acid methyl-transferase activity have an opposite impact on the efficiency of industrial kraft pulping. *Plant Physiol.* 119: 153-163.

Lazo G.R., S.P.A., Ludwig R.A. 1991. A DNA transformation-competent Arabidopsis genomic library in Agrobacterium. *Biotechnology* 963-967.

Lees, G.L. 1992. Condensed tannins in some forage legumes: their role in the prevention of ruminant pasture bloat. *Basic Life Sci.* 59: 915–934.

Li, Y.G., Tanner, G., Larkin, P. 1996. The DMACA-HCl protocol and the threshold proanthocyanidin content for bloat safety in forage legumes. *J. Sci. Food Agric.* 70: 89-101.

Li, H., Flachowsky, H., Fischer, T.C., Hanke, M.V., Forkmann, G., Treutter, D., Schwab, W., Hoffmann, T. and Szankowski, I. 2007. Maize *Lc* transcription factor enhances biosynthesis of anthocyanins, distinct proanthocyanidins and phenylpropanoids in apple (*Malus domestica* Borkh.). *Planta* 226: 1243-1254.

Li, P., Chen, B., Zhang, G., Chen, L., Dong, Q., Wen, J., Mysore, K.S. and Zhao, J. 2016. Regulation of anthocyanin and proanthocyanidin biosynthesis by *Medicago truncatula* bHLH transcription factor *MtTT8*. *New Phytologist*, doi:10.1111./nph.13816.

Liu, Ch., Jun, J.H. and Dixon, R.A. 2014. *MYB5* and *MYB14* play pivotal roles in seed coat polymer biosynthesis in *Medicago truncatula*. *Plant Physiol.* 165: 1424-1439.

Lloyd, A.M., Walbot, V. and Davis, R.W. 1992. Arabidopsis and Nicotiana anthocyanin production activated by maize regulators R and C1. *Science* 258: 1773-1775.

Ma, J.K., Hiatt, A., Hein, M., Vine, N.D., Wang, F., Stabila, P., van Dolleweerd, C., Mostov, K. and Lehner, T. 1995. Generation and assembly of secretory antibodies in plants. *Science* 268: 716–719.

McMahon, L.R., McAllister, T.A., Berg, B.P., Majak, W., Acharya, S.N., Popp, J.D., Coulman, B.E., Wang, Y. and Cheng, K.J. 2000. A review of the effects of forage condensed tannins on ruminal fermentation and bloat in grazing cattle. *Can J Plant Sci* 80: 469-485.

Muir, A.D., Gruber, M.Y., Hinks, C.F., Lees, G.L., Onyilagha, J., Hallett, R., Xia, F., Soroka, J. and Erlandson, M. 1999. The effects of condensed tannin in the diets of major crop insects. In GG Gross, RW Hemingway, T Yoshida, eds, Plant Polyphenols 2: Chemistry, Biology, Pharmacology, Ecology. Plenum Press, New York, pp. 867-881.

Naqvi, S., Farré, G., Sanahuja, G., Capell, T., Zhu, C. and Christou, P. 2009. When more is better: multigene engineering in plants. *Trends in Plant Science* 15 (1): 48-56.

Niezen, K.E., Waghorn, T.S., Charleston, W.A.C. and Waghorn, G.C. 1995. Growth and gastrointestinal nematode parasitism in lambs grazing either Lucerne (*Medicago sativa*) or sulla (*Hedysarum coronarium*) which contains condensed tannins. *J Agric Sci* 125: 281-289.

Niezen, J.H., Robertson, H.A., Waghorn, G.C. and Charleston W.A.C. 1998. Production, faecal egg counts and worm burdens of ewe lambs which grazed six contrasting forages. *Veterin Parasitol* 80: 15-27.

Pang, Y., Peel, G.J., Wright, E., Wang, Z. and Dixon, R.A. 2007. Early steps in proanthocyanidin biosynthesis in the model legume *Medicago truncatula*. *Plant Physiol.* 145: 601-615.

Park, K.I., Ishikawa, N. Morita, Y., Choi, J.D., Hoshino, A. and Lida, S. 2007. A bHLH regulatory gene in the common morning glory, *Ipomoea purpurea*, controls anthocyanin biosynthesis in flowers, proanthocyanidin and phytomelanin pigmentation in seeds and trichome f ormation. *Plant J.* 49: 641-654.

Peel, G.J., Pang, Y., Modolo, L.V. and Dixon, R.A. 2009. The *LAP1* MYB transcription factor orchestrates anthocyanidin biosynthesis and glycosylation in Medicago. *Plant J.* 59: 136- 149.

Polturak, G., Breitel, D., Grossman, N., Sarrion-Perdigones, A., Weithorn, E., Pliner, M., Orzaez, D., Granell, A., Rogachev, I. and Aharoni, A. 2016. Elucidation of the first committed step in betalain biosynthesis enables the heterologous engineering of betalain pigments in plants. *New Phytol.* 210 (1): 269-83.

Qi, B., Fraser, T., Mugford, S., Dobson, G., Sayanova, O., Butler, J., Napier, J.A., Stobart, A.K. and Lazarus, C.M.(2004. Production of very long chain polyunsaturated omega-3 and omega-6 fatty acids in plants. *Nat. Biotechnol.* 22: 739-745.

Ramsay, N.A., Walker, A.R., Mooney, M. and Gray, J.C. 2003. Two basic helix-loop-helix genes (*MYC-146* and *GL3*) from Arabidopsis can activate anthocyanin biosynthesis in a white flowered *Matthiola incana* mutant. *Plant Mol. Biol.* 52: 679-688.

Robbins, M.P., Paolocci, F., Hughes, J.-W., Turchetti, V., Allison, G., Arcioni, S., Morris, P. and Damiani, F. 2003. *Sn*, a maize bHLH gene, modulates anthocyanin and condensed tannin pathways in *Lotus corniculatus*. *J. Exp. Bot.* 54: 239-248.

Roberts, C., Rajagopal, S., Smith, L.M., Nguyen, T.A., Yang, W., Nugrohu, S., Ravi, K.S., Vijayachandra, K., Harcourt, R.L., Dransfield, L., et al. 1997. A Comprehensive Set of Modular Vectors for advanced Manipulations and Efficient Transformation of Plants. pCAMBIAVector Release Manual Rockefeller Foundation Meeting of the International Program on Rice Biotechnology, September 15-19, Malacca, Malaysia.

Ruiz-Lopez, N., Haslam, R.P., Venegas-Caleron, M., Li, T., Bauer, J., Napier, J.A., Sayanova, O. 2012. Enhancing the accumulation of omega-3 long chain polyunsaturated fatty acids in transgenic *Arabidopsis thaliana* via iterative metabolic engineering and genetic crossing. *Transgenic Res.* 21: 1233-1243.

Samac, D.A., Tesfaye, M., Dornbusch, M., Saruul, P. and Temple, S.J. 2004. A comparison of constitutive promoters for expression of transgenes in alfalfa (*Medicago sativa*). *Transgenic Res*, 13, 349-361.

Samac, D.A. and Austin-Phillips, S. 2006. Alfalfa (*Medicago sativa* L.). *Methods Mol Biol*, 343, 301-311.

Sarrion-Perdigones, A., Falconi, E.E., Zandalinas, S.I., Juarez, P., Fernandez-del-Carmen, A., Granell, A.and Orzaez, D. 2011. GoldenBraid: An Iterative Cloning System for Standardized Assembly of Reusable Genetic Modules. *PLoS ONE* 6 (7) e21622.

Sarrion-Perdigones, A., Vázquez-Vilar, M., Palací, J., Castelijns, B., Forment, J., Ziarsolo, P., Blanca, J., Granell, A. and Orzaez, D. 2013. GOLDENBRAID 2.0: A comprehensive DNA assembly framework for Plant Synthetic Biology. *Plant Physiol.* 162(3): 1618-31.

Sarrion-Perdigones, A., Palaci, J., Granell, A. and Orzaez, D. 2014. Design and Construction of Multigenic Constructs for Plant Biotechnology Using the GoldenBraid Cloning Strategy. *DNA Cloning and Assembly Methods, Methods in Molecular Biology*, vol. 1116, doi 10.1007/978-1-62703-764-8_10.

Schwinn, K., Venail, J., Shang, Y.J., Mackay, S., Alm, V., Butelli, E., Oyama, R., Bailey, P., Davies, K. and Martin, C. 2006. A small family of MYB-regulatory genes controls floral pigmentation intensity and patterning in the genus Antirrhinum. *Plant Cell* 18: 831-851.

Sweeney, M.T., Thomson, M.J., Pfeil, B.E. and McCouch, S. (2006). Caught redhanded: Rc encodes a basic helix-loop-helix protein conditioning red pericarp in rice. *Plant Cell* 18: 283-294.

Tanner, G.J., Francki, K.T., Abrahams, S., Watson, J.M., Larkin, P.J. and Ashton, A.R. 2003. Proanthocyanidin biosynthesis in plants. Purification of legume leucoanthocyanidin reductase and molecular cloning of its cDNA. *J. Biol.* Chem. 278: 31647-31656.

Trinh, T.H., Ratet, P., Kondorosi, E., Durand, P., Kamat, K., Bauer, P. and Kondorosi, A. 1998. Rapid and efficient transformation of diploid *Medicago truncatula* and *Medicago sativa* ssp. falcata lines improved in somatic embryogenesis. *Plant Cell Reports*, 17: 345-355.

Twyman, R.M., Christou, P. and Stoger, E. 2002. Genetic transformation of plants and their cells. In Plant Biotechnology and Transgenic Plants (Oksman-Caldentey, K.M. and Barz, W., eds), pp. 111-141, Marcel Dekker Inc.

Vafaee, Y., Staniek, A., Mancheno-Solano, M. and Warzecha, H. 2014. A Modular Cloning Toolbox for the Generation of Chloroplast Transformation Vectors. Isalan M, ed. *PLoS ONE* 9 (10): e110222.

Verdier, J., Zhao, J., Torres-Jerez, I., Ge, S., Liu, Ch., He, X., Mysore, K.S., Dixon, R.A. and Udvardi, M.K. 2012. *MtPAR* MYB transcription factor acts as an on switch for proanthocyanidin biosynthesis in *Medicago truncatula. Proc Natl Acad Sci U S A* 109 (5): 1766-1771.

Waghorn, G.C. 1990. Beneficial effects of low concentrations of condensed tannins in forages fed to ruminants. In DE Akin, LLG Ljungdahl, JR Wilson, PJ Harris, eds, Microbial and Plant Opportunities to Improve Lignocellulose Utilization by Ruminants. Elsevier, New York, pp. 137-147.

Xie, D.Y., Sharma, S.B., Paiva, N.L., Ferreira, D. and Dixon, R.A. 2003. Role of anthocyanidin reductase encoded by *BANYULS* in plant flavonoid biosynthesis. *Science* 299: 396-399.

Xie, D.Y., Sharma, S.B. and Dixon, R.A. 2004. Anthocyanidin reductases from *Medicago truncatula* and *Arabidopsis thaliana*. *Arch. Biochem. Biophys.* 422: 91-102.

Xie, D.Y., Sharma, S.B., Wright, E., Wang, Z.-Y. and Dixon, R.A. 2006. Engineering plants for introduction of health-beneficial proanthocyanidins. *Plant J.* 45: 895-907.

In: Legumes for Global Food Security ISBN: 978-1-53612-265-7
Editors: Alfonso Clemente et al. © 2017 Nova Science Publishers, Inc.

Chapter 9

LEGUME ALLERGENICITY: THE EFFECT OF FOOD PROCESSING

Carmen Cuadrado * ***and Mercedes M. Pedrosa***
Departamento de Tecnología de Alimentos,
SGIT-INIA, Madrid, Spain

ABSTRACT

Legume proteins are important food ingredients because of their nutritional and nutraceutical properties. However, legumes are also considered relevant in the development of food allergies through ingestion. Peanuts and soybeans are important food allergens in the Western countries, while lentil, chickpea and lupin allergens are more relevant in the Mediterranean area. Moreover, lupin seed flour is being increasingly used in bakery, snacks and pastry products due to its multifunctional properties. Food processing, such as the enzymatic digestion and the treatments combining heat and steam pressure, i.e., autoclave and Instant Controlled Pressure Drop (DIC®), can decrease legume allergenicity. This chapter reviews the effect of several processing conditions on legumes such as lupin, peanut, lentil, chickpea and soybean by SDS-PAGE, immunoblotting and ELISA. The data

* Corresponding Author Email: cuadrado@inia.es.

demonstrate that food processing produces a decrease in the overall in vitro IgE binding capacity of peanut, lentil and chickpea and a drastic reduction of lupin and soybean immunoreactivity.

Keywords: allergens, food hypersensitivity, legumes, thermal processing, pressure processing, enzymatic digestion

1. INTRODUCTION

In the past few decades, the prevalence of allergic diseases has increased dramatically and food allergies play a major role in this increase. Food allergy affects approximately to 6-8% children and 3-4% adults in Europe (Mills et al. 2007, Chung and Champagne 2001, Beyer et al. 2001, Chung S. Y. and E. T. 1999, Chung SY and ET 2001, Chung and Champagne 1999). Legumes are increasingly being regarded as beneficial food ingredients. In fact, they are recommended as staple food by health organizations and dieticians are nowadays tending to encourage their consumption in counseling (Leterme 2002). Allergenic potential of many legumes as lupins (*Lupinus albus*), peanuts (*Arachis hypogaea*), lentils (*Lens culinaris*), chickpeas (*Cicer arietinum*) and soybeans (*Glycine max*) is well established (Verma et al. 2013). As legume allergy affects to a significant proportion of population, some effective methods should be adopted to minimize its risk. Consumption of legume may provoke mild to severe anaphylatic symptoms in sensitized individuals. The usual symptoms are angioedema, vomiting, urticaria, allergic rhinitis, diarrhea, rashes on skin, swelling of the tongue or throat and asthma. Moreover, legumes have significant degree of immunological cross-reactivity among themselves and with other plant allergens. Therefore, there is a growing interest in the development of newer methods to overcome such type of allergic problems and several strategies are undergoing for minimizing allergenicity potential of legumes crops (Chung and Reed 2011).

Heating promotes protein denaturation, aggregation, and structure disruption and therefore, it can modify allergenic properties of proteins.

The molecular basis of changes in allergenic activity is the inactivation or destruction of epitope structures, the formation of new epitopes, or an enhanced access to cryptic epitopes by denaturation of the native allergen (Besler, Steinhart, and Paschke 2001). Plant protein allergenicity may be variably affected by thermal processing, i.e., increasing or decreasing IgE immunoreactivity. Thus, the overall effect of such procedures on the complex food matrix cannot be predicted (Beyer et al. 2001, Mondoulet et al. 2005). A better understanding of how thermal processing induced biochemical and immunological changes in food allergens may help to develop new diagnostic tools and to alleviate the allergenic problem.

Our previous studies have observed that boiling in autoclave at harsh conditions reduced lupin, lentil, chickpea, soybean and peanut allergenicity drastically (Álvarez-Álvarez et al. 2005, Cabanillas, Maleki, et al. 2012, Cuadrado et al. 2007, Cuadrado et al. 2009). However, several extremely resistant immunoreactive proteins still remained in some legumes, as lentil and chickpea, even after these extreme treatments. Similar results have been found using Instant Controlled Pressure-Drop (DIC®) treatment, a procedure that combines heat and steam pressure as in autoclaving (Cuadrado et al. 2011, Guillamón et al. 2008).

A very effective way to reduce or remove allergenicity is enzymatic hydrolysis, destroying the responsible epitopes. By this way, hypoallergenic foods can be developed. The downside of this treatment is, of course, that the properties of the resulting food are drastically different from the original food. The treatment notably affects texture and taste, as well as appearance. The resistance to enzymatic hydrolysis is, incidentally, also of importance for the ocurrence of allergic reaction upon digestion (Wickham, Faulks, and Mills 2009). Two studies have analyzed the effects of the sequential action of an endoprotease from *Bacillus licheniformis* (Alcalase) and an exoprotease from *Aspergillus oryzae* (Flavourzyme) in peanut (Cabanillas, Pedrosa, et al. 2012) and lentil allergenicity (Cabanillas et al. 2010). These enzymes produce protein hydrolysates with a significant higher degree of hydrolysis than papain, trypsin, and α-chymotrypsin (Humiski and Aluko 2007).

Table 1. Allergens in Legumes

Source	Allergen	Protein family	MW (kDa)	References
Peanut	Ara h1	Vicilin (7S vicilin like globulin)	64	Burks et al. 1991
(*Arachis*	Ara h2	Conglutin (2S albumin)	17	Burks et al. 1992
hypogea	Ara h3	Cupin (Legumin type 11S,	60	Jin et al. 2009
	Ara h4	Glycinin)	37	Kleber-Janke et al.
	Ara h5	Cupin (Legumin type 11S,	15	1999
	Ara h6	Glycinin) Profilin	15	Kleber-Janke et al.
	Ara h7	Conglutin (2S albumin)	15	2001
	Ara h8	Conglutin (2S albumin)	17	Van Wijk et al. 2004
	Ara h9	Pathogen related protein (PR-10)	9.8	Kleber-Janke et al.
	Ara h10	Non-specific lipid transfer protein 1	16	1999
	Ara h	18 Kda oleosin	14	Mittag et al. 2004
	Agglutinin	14 Kda oleosin		Asero et al. 2000
				Pons et al. 2002
				Burks et al. 1994
Soy	Gly m1	Non specific lipid transfer protein	7	Helm et al. 1998
(*Glycine*	Gly m2	Defensin	8	Wilson et al. 2005
max)	Gly m3	Profilin	14	Ogawa et al. 2000
	Gly m4	Pathogen related protein (PR-10)	17	Renaud et al. 1996
	Gly m Bd28K	Cupin (7S vicilin like globulin)	28	Gonzalez et al. 1992
	Gly m Bd30K	Thiol protease of the papain	34	Gonzalez et al. 1992
	Gly m Lectin	superfamily Lectin, SBA	14.5	Mittag et al. 2006
	Gly m Bd 60 K	(agglutinin)	63–67	Codina et al. 1999
	Gly m glycinin	Cupin(7S vicilin like globulin)	40	Natarajan et al. 2006
	G1	Glycinin, 11S seed storage protein	22	Natarajan et al. 2006
	Gly m glycinin	Glycinin, 11S seed storage protein	61.2	Natarajan et al. 2006
	G2	Glycinin, 11S seed storage protein	20	Burks et al. 1994
	Gly m glycinin	Kunitz trypsin inhibitor		
	G4			
	Gly m TI			
Lentil	Len c1	Vicilin	48	Sanchez-Monge et al.
(*Lens*	Len c 2	Seed biotinylated protein	66	2000 Sanchez- Monge
culinaris)	Len c 3	Non specific lipid transfer protein 1	9	et al. 2000 Akkerdas
				et al. 2012
Chickpea	Cic a 2S	Albumin Prolamin (2S albumin)	10–12	Vioque et al. 1999
(*Cicer*	Cic a 11S	Cupin (Legumin type 11S,	46.5 and	Karamloo et al. 2001
arietinum)	Globulin	Glycinin)	39.8 kDa	
			(Acidic	
			peptide)	
			25.3 and	
			24.3 kDa	
			(Basic	
			peptide)	
Lupin	Lup an1	Conglutin beta (7S seed storage	55–61	Goggin et al. 2008
(*Lupinus*	Lup a vicilin	globulin, vicilin)	34.5	Guillamon et al. 2010
sp)	Lup a 11S	Cupin	20	Guillamon et al/ 2010
	globulin	Cupin		

In the present chapter, we have summarized the impact of thermal and non thermal processing on the IgE binding capacity of proteins from legumes such as lupin, peanut, lentil, chickpea and soybean.

2. Major Legumes and Their Predominant Allergens

Legumes are a rich source of protein and other nutritious elements with functional properties, which promote their presence in manufactured foods. Therefore, legumes have become an important ingredient of human diet throughout the world (Duranti 2006). But unfortunately, legumes are one of the common food with potential to elicit allergenic response. Some common allergenic legumes are peanut, soybean, lentil, lupin, pea, chickpea, red gram, and blackgram, etc. Several legume allergens have been identified and characterized as belongings to the cupin superfamily of storage proteins (Mills et al. 2002). The Table 1 summarize the main legume allergens. The major allergens of soybean and peanut have been extensively studied and Ara h 1, a 65 kDa glycoprotein belonging to the vicilin family, is one of the best characterized (Burks, Sampson, and Bannon 1998). Lentil and chickpea allergens have been less investigated. Two chickpea allergens have been identified and only three lentil allergens have been characterized until now. Researching is going on by various groups for identification of lupin allergens (Guillamón, Rodríguez, et al. 2010, Verma et al. 2013). Several legume allergens have shown high resistance to digestion and food processing, thus strengthening their allergenic potential (Mills et al. 2004).

The pattern of legume sensitization varies in different parts of the world, probably due to the genetic status of individuals, consumption habits or may be involvement of other factors. Higher prevalence of peanut allergy is found in the United Kingdom, France, and North America whereas major incidence of soybean allergy is reported in south-east Asia. Lentil, chickpea and lupin allergic reactions are more important in the

Mediterranean area (Crespo and Rodriguez 2003). According to UE regulations it is mandatory to label a list of 14 groups of potential allergenic foods, among which, peanut and soybean were firstly included, and lately, lupin as well as its derived ingredients (Commission Directive 2006/142/EC).

3. Food Processing

Consumers buy processed foods for a variety of reasons, being convenience perhaps the most important one. Additional reasons for buying processed foods include variety, out of season availability of particular food in processed form (e.g., several fruits and vegetables), lack of time/skills/equipment needed for home food preparation of certain foods, and assured consistent quality and safety. Regardless of the type of food or the processing method to which a food is subjected to, whether at home, institutional, or industrial setting; the reasons why foods are processed include: (i) improvement of food qualities such as flavor, texture, taste, color; (ii) improve preservation and safety; (iii) enhance suitability for specific product applications; (iv) convenience, pleasure, and variety; (v) obtain or generate useful by-products; and (vi) increase marketability and/or revenue. In addition to industrial food processing, a significant quantity of food is processed by consumers at home and in institutional settings. The choice of processing method can be influenced by available infrastructure, product type, scale of processing, consumer preference, product sensory qualities, and economics. For these and other reasons, the same raw food is often processed differently in different markets.

Regardless of the product, the scale of processing, and desired end-product qualities, food processing is based on a few basic principles. For example, from a food preservation point of view, it is desirable to control water activity (aw) of a food product. Alternatively, food products can be treated with methods or substances that inhibit growth and/or viability of microorganisms (and their spores). Food being a live matter contains a variety of chemical and biochemical components that affect product quality positively or negatively. Thus, careful control of interactions between food components is essential in managing food quality, and inactivation of microorganisms to protect food from microbial spoilage may not be sufficient to produce and preserve an acceptable final product. Additional measures may be required, such as use of food additives and preservatives, to prevent unwanted chemical and biochemical interactions that occur in the food product. Most food processing methods involve energy input and/ or mass transfer. During processing, the processed food may remain in a single phase (state) or may experience phase change. For example, during pasteurization, milk remains in one phase (liquid) while cheese manufacture involves phase change as milk (liquid phase) is converted to curd (milk protein curd in solid phase) and whey (liquid). Similarly, the heating medium (e.g., steam) may experience a phase change (steam- vapor phase condensing to hot water- liquid phase). Such phase changes are often accompanied by large energy transfers. A variety of processes are used to tailor the demands of processed food product and economy. As a result, myriad of food processing methods are used. Despite the variety, food processing methods can be broadly classified as thermal (e.g., blanching, canning, cooking, and steaming) and nonthermal (e.g., washing, filtration, enzymatic modification, irradiation, high pressure processing). In some cases, both thermal and nonthermal processing methods are used to produce a single food product (e.g., canned vegetables and fruits).

4. Effect of Thermal Processing on Legume Allergens

Foods are subjected to thermal and non thermal processing methods to improve their quality, preservation, safety and suitability for specific product applications. The degree of processing affects the digestibility, solubility and other parameters related. During processing, proteins can form oligomers, become denatured, aggregated, fragmented and re-assembled and these changes most often reduce solubility (Maleki 2004). Processing can alter the overall IgE binding profiles of a particular extract becoming more o less antigenic or resulting in new allergens (neoallergens) (Schmitt et al. 2010). Therefore, the study of processing is necessary for assessing the allergenicity of existing and newly introduced foods (Wal 2003). The effect of thermal processing mainly depends on temperature and time used. Moreover, alteration in structure of protein also depends on interactions with other food matrix constituents. In general, when temperature is around 70-80°C, loss of secondary structure occurs whereas at 80-90°C, formation of new bonds and rearrangements of disulfide bonds takes place. At higher temperature (90-100°C) aggregate formation appears (Davis and Williams 1998).

Studies with peanut (Maleki et al. 2000a, Chung et al. 2003) demonstrated that extracts from roasted bind IgE at approximately 90-fold higher levels than raw peanuts. These studies demonstrated that the major allergens, Ara h 1 and Ara h 2, from roasted peanuts have undergone structural alterations that enhance their allergenic properties. Heat treatment probably increases the digestibility of protein, so the absorption of protein in gastrointestinal tract can also increase, and because of this, the possibility of allergenic protein to elicit an allergenic response decreases. However, in some cases, thermal processing may reduce the digestibility of a particular allergen or some neoantigens that were not originally present may be formed. This global phenomenon may enhance the allergenic problem in sensitized patients and also these neoantigens may present an additional problem. One major factor responsible for the formation of some neoantigens is the Maillard reaction i.e., the interaction of protein component with sugar residues upon heating, generating sugar conjugated protein derivatives, which enhance the allergenicity of protein (Maleki et al. 2000a).

The IgE antibodies recognize and interact with epitopes present on allergenic proteins. Two types of IgE binding epitopes are possible, either linear or conformational ones. In linear epitopes, amino acid are arranged in linear order in the polypeptide chain while in the conformational, amino acids that are far apart in the primary sequence but may come together during folding of polypeptide chain. Linear epitopes may be more problematic as compared to the conformational ones as these are mostly resistant to heat treatment. Thermal processing mainly affects conformational epitopes as the bonds can be broken down due to heat. Refolding allows formation of native conformational epitopes but few new allergens may be formed which needs further efforts to minimize the risk associated with neoantigen (Sathe and Sharma 2009). Thus, thermal processing as well as other processing events can dramatically alter the structure, function and allergenicity of foods. The incomplete knowledge of the allergens from processed foods increases the complexity of food allergies diagnosis.

5. Effect of Enzymatic Digestion on Legume Allergens

A number of studies have evaluated the effects of enzymatic hydrolysis on the allergenicity and digestibility of food proteins. Cow's milk hydrolysate formulae have been developed to decrease or eliminate its allergenicity (Terracciano et al. 2002, Merritt et al. 1990). Changes in allergenicity of enzymatically hydrolyzed legumes have been evaluated mainly in soybean (Lee et al. 2007, Burks et al. 1992) and peanut (Burks et al. 1992). Studies with other members of the Leguminosae family are scarce. Clemente et al. (1999) evaluated the reduction of immunoreactivity to chickpea allergens during extensive hydrolysis generated by the sequential action of an endoprotease (Alcalase) and an exoprotease (Flavourzyme). A decrease in IgE-binding capacity by more than 90% was detected using sera from six patients who reported reactions after chickpea ingestion. Moreover, this sequential hydrolysate is considered an interesting material as food additive due to its high protein quality (Clemente, Vioque, Sanchez-Vioque, Pedroche, Bautista, et al. 1999). Sormus de Castro Pinto et al. (2009) investigated the effects of pepsin and trypsin on the in vitro reactivity of 11S globulin of lupine, chickpea, and lentil using rabbit antiserum for these proteins. They obtained a large amount of short peptides and free amino acids after enzymatic treatment. The assessment of different enzyme activities to decrease allergenicity seems to be important; moreover, these studies should be carried out with sera from patients with documented clinical allergy to the source food (Lucas and Atkinson 2008). Enzymatic protein hydrolysates constitute an alternative to intact proteins in the development of special formulations designed to provide nutritional support to specific population groups with different needs, such as infants, elderly, and food-allergic patients. In addition, protein hydrolysates show technological advantages. The combination of extensive enzymatic treatment coupled with post hydrolysis food processing procedures, such as heat treatment, ultrafiltration, etc., is considered the most effective way to obtain protein products with an added high value for human nutrition and reduced allergenicity (Clemente 2000).

6. Effect of Processing on Lupin (*Lupinus albus*) Allergens

White lupin is considered to be a rich source of protein with notable content of lysine and is being increasingly used due to its multifunctional properties and its potential hypocholesterolemic and hypoglycaemic effects (Duranti 2006). However, lupin seed flour has been reported as a causative agent of allergic reactions, especially in patients with allergy to peanut (Moneret-Vautrin et al. 1999). Moreover, lupin seed proteins have probed to be an interesting model to study protein thermal conformational stability under different pH conditions from both, biochemical and techno-functional, points of view (Duranti et al. 2000).

In a previous study (Álvarez-Álvarez et al. 2005) microwave cooking, boiling and extrusion cooking produced minimal changes on IgE binding to lupin proteins, however, boiling in autoclave at 2.6 bar (138°C) for 20 min produces relevant decrease in allergenicity. Only two proteins, 23 and 29 kDa, have IgE-binding capacity after autoclaving for 20 min, whereas autoclaving for 30 min completely abolished IgE binding of these components, suggesting that lupin allergens are relatively heat stable. Later on, two main allergens were identified as Lup 1 (conglutin β, 34.5 kDa, 7S protein) and Lup 2 (a basic subunit of conglutin α, 20 kDa, 11S protein) (Guillamón, Rodríguez, et al. 2010). Both were partially sequenced and their high homology degree with major allergens from peanut, lentil, pea and soybean could explain the IgE cross-reactivity of lupin with these legumes.

Guillamón et al.(2008) studied the effect of DIC treatment on lupin allergenicity. Raw and DIC treated lupin at 3 bar for 1 and 3 min have similar SDS and IgE-immunoblot band pattern. In these samples, Lup 1 (34.5 kDa) and Lup 2 (20 kDa) were still present. Similar protein pattern was found by Álvarez-Álvarez et al. (2005) in autoclaved lupin seeds at 1.2 bar for 20 min. A mayor decrease in the number and intensity of bands was observed when DIC was applied at 6 bar for 1 min, but Lup 1 (34.5 kDa) was still present. However, after 3 min at 6 bar nor Lup 1 neither Lup 2 were detected.This study confirmed the thermal resistance of the major lupin allergens previously reported from autoclave treatments (Álvarez-Álvarez et al. 2005). Both results suggest that lupin seeds treated with processing technologies combining heat and steam pressure, such as DIC and autoclave, could almost eliminate lupin in vitro immunoreactivity. Moreover, the breadmaking properties of this hypoallergenic lupin flours have been studied (Guillamón, Cuadrado, et al. 2010) and the results indicated that thermally-treated lupin flours, had similar breadmaking and sensorial properties as untreated lupin flour. These thermal treatments could increase the potential use of lupin flour as a food ingredient while reducing the risk to provoke allergic reactions.

7. Effect of Processing on Peanut (*Arachys hipogaea*) Allergens

Peanut allergy is one of the most common IgE-mediated reactivity to food because of its severity and lifelong persistence (Sicherer and Sampson 2007). Considerable effort has been spent in characterizing peanut allergens and eleven allergenic proteins have been identified until now (Ara h 1-Ara h 11). The major peanut allergens, Ara h 1 (65 kDa, vicilin) and Ara h 2 (17 kDa, conglutin), are recognized by 70-90% of sensitized subjects (Burks, Sampson, and Bannon 1998) and Ara h 3 (11 S legumin) has been considered to play a lesser allergenic role (Rabjohn et al. 1999). The thermal treatment has significant effect on peanut immunoreactivity. As previously mentioned, roasting peanut enhances its IgE-binding capacity (Maleki et al. 2000a), while boiling decreases its allergenicity (Beyer et al. 2001).

According to Cabanillas et al. (2012) IgE immunoreactivity of roasted peanut decreased significantly at extreme conditions of autoclaving (2.6 bar, 30 min). The results obtained by circular dichroism spectroscopy indicated that most of the α-helical structure was lost after that autoclave treatment. The fact that many of the IgE binding epitopes of peanut major allergens (Ara h 1, Ara h 2 and Ara h 3) were located on the α-helical regions can explain such decrease (Barre et al. 2007, Mueller et al. 2011).

When DIC treatment was applied to raw and roasted peanut proteins did not produce any relevant change in the immunoblot profile at 3 bars for 1 and 3 min and after DIC treatment at 6 bar for 1 min compared to untreated samples (Cuadrado et al. 2011). However, DIC treatment at 6 bars for 3 min resulted into a remarkable decrease in the protein bands of 65 kDa (putative Ara h 1) and no immunoreactive bands lower than 20 kDa were recognized. However, some bands (e.g., 37 kDa, Ara h 3 acid subunit) behave differently in raw and roasted peanuts. Taking into account the reduction of in vitro immunoreactivity observed at the highest pressure and time applied, the DIC technique seems to be more effective on roasted peanut proteins immunoreactivity than on raw peanuts. From the literature, it is apparent that immunoreactivity changes of peanut proteins following thermal treatment may be largely due to modification in structure as well as reactivity of each individual allergen of peanut and also their interaction with the food matrix, but more studies are still required to open a broad spectrum of our understanding regarding this concern.

Several studies have evaluated the effects of enzymatic hydrolysis on the allergenicity and digestibility of peanut proteins in order to explore strategies designed to decrease the ability of peanuts to elicit dangerous allergic responses. The effect of the individual and sequential action of Alcalase and Flavourzyme in the IgE-binding properties of the soluble protein fraction of roasted peanut (RP), using sera from patients with a clinical allergy to peanut has been investigated by Cabanillas et al. (2012). In this study an endoprotease (Alcalase) and an exoprotease (Flavourzyme) were used, both individually and sequentially to evaluate their effects in the IgE antibody reactivity to RP protein extract. RP was selected in this study because it has been recognized that it is more allergenic than raw peanut (Maleki et al. 2000b, Beyer et al. 2001, Maleki et al. 2003, Chung and Champagne 1999, Chung and Champagne 2001). RP extract binds IgE from patients with peanut allergy at approximately 90-fold higher levels than that of raw peanuts, and the protein modifications induced by the Maillard reaction contribute to the observed effect (Maleki et al. 2000b, Chung and Champagne 2001, Chung and Champagne 1999). The results showed that individual treatment with Flavourzyme at 300 min reached a higher hydrolysis degree (DH) than individual treatment with Alcalase at 150 min; however, SDS-PAGE protein patterns were very different in both cases. In contrast to Flavourzyme hydrolysates, Alcalase hydrolysates showed no detectable stained proteins in SDS-PAGE. Immunoblot and ELISA assays carried out with the serum pool from the 5 patients with a clinical allergy to peanuts showed a decrease of IgE reactivity and after 30 min of hydrolysis, no bands were detected in immunoblotting and a 100% reduction in IgE reactivity was observed in ELISA. These results were confirmed by Western blot with individual sera. None of the sera recognized any RP proteins after sequential endo- and exoprotease hydrolysis. In contrast, individual treatment with Flavourzyme caused an increase in IgE reactivity detected by ELISA at 30 min. However, Flavourzyme led to a 65% decrease in IgE reactivity at the end of the assay (300 min). These results show that hydrolysis with Alcalase produces a decrease in IgE recognition in RP protein extract due to a decrease in the main peanut allergens. These enzymatically treated protein hydrolysates could constitute an alternative to intact proteins in the development of different products.

8. Effect of Processing on Lentil (*Lens culinaris*) Allergens

Lentil is commonly consumed in Mediterranean areas and it has been reported as a cause of IgE-mediated hypersensitivity reactions, particularly in pediatric patients. Multiple IgE binding allergens have been detected in both raw and boiled lentil but the studies investigating the allergenicity of lentils with well documented clinical sera are scarce. Two major lentil allergens Len c 1 (48 kDa vicilin) (Lopez-Torrejon et al. 2003) and Len c 2 (66 kDa) have been characterized and isolated from boiled lentils (Sanchez-Monge et al. 2000). More recently, a third allergen, Len c 3, has also been characterized (Akkerdaas et al. 2012).

The effect of boiling on lentil proteins does not seem to be effective in reducing its allergenic potential. However, autoclave treatment of lentil (1.2 and 2.6 bar, up to 30 min) significantly decrease the IgE binding proteins to a great extent. Autoclaving at the highest pressure (2.6 bar) for 30 min has a prominent effect on minimizing the global IgE-binding capacity, although extremely resistant immunoreactive proteins still remained in this legume even after this harsh process (Cuadrado et al. 2009).

When we study the effect of DIC-treatment on lentil (Cuadrado et al. 2011) observed that the control sample was composed of numerous immunoreactive bands with molecular weights between 101 and 140 kDa, included a protein of 48 kDa (putative major allergen Len c 1). DIC processing at extreme conditions (6 bar, 3 min) produced a reduction of the number of IgE-binding proteins similar to that observed with autoclave at harsh conditions. After such DIC-treatment only four heat-stable allergenic proteins of MW 29, 48, 57 ad 68 kDa were still present. These results confirmed the thermostability of some lentil allergens also detected after autoclave processing.

There are only a few studies carried out to evaluate the effect on IgE reactivity of the food-hydrolysis products with sera from patients with well-documented allergy to these foods. In the study reported by Cabanillas et al. (2010), lentil protein extract was hydrolyzed by sequential action of an endoprotease (Alcalase) and an exoprotease (Flavourzyme). Clemente et al. (1999) obtained the most effective reduction of antigenicity in chickpea by a sequential treatment with Alcalase and Flavourzyme compared with both enzymes tested individually. The results showed in sequential lentil hydrolysis, a 24% DH at 180 min produced by Alcalase; the addition of Flavourzyme, increased the DH to 56% in the final reaction time (480 min). These data seem to indicate that endoprotease Alcalase produced an initial digestion that increased the number of target sites for the action of the exoprotease Flavourzyme. These values are similar to those obtained for pea (Humiski and Aluko 2007) and soy (Seo, Lee, and Baek 2008) protein hydrolysates. Immunoreactivity to raw and hydrolyzed lentil extract was evaluated by means of IgE immunoblotting and ELISA using sera from five patients with clinical allergy to lentil. The results indicated that sequential hydrolysis of lentil results in an important proteolytic destruction of IgE-binding epitopes shown by in vitro experiments. However, some allergenic proteins were still detected by sera from four out of five patients in the last step of sequential hydrolyzation.

9. Effect of Processing on Chickpeas (*Cicer arietinum*) Allergens

Chickpea is an important source of proteins in several parts of the world. In Asian countries it is widely consumed in many traditional dishes. The high consumption rate of this crop in the Mediterranean area has also resulted in allergic problem in sensitive individuals (Crespo and Rodriguez 2003). Chickpea and lentil are the most frequent cause of allergic reactions to legumes in Spanish children and there exist a cross reactivity between both (Crespo et al. 1995). Some subjects allergic to this legume on ingestion also report symptoms when they inhale vapors from cooking chickpeas (Niphadkar, Patil, and Bapat 1997). Previous studies have detected the multiple IgE-binding bands in chickpea extracts in the molecular weight range of 10–106 kDa, of which the majority were found to be heat stable (Patil, Niphadkar, and Bapat 2001). So far, allergens of chickpea have not been immunologically characterized, only two allergenic polypeptides (2S albumin and 11S globulin) from chickpea have been identified (Vioque et al. 1999).

Cuadrado et al. (2009) found multiple IgE-binding proteins in chickpea boiled for 30 min. A decrease in the number and intensity of bands was observed after autoclaving (1.2 bar, 12 min) and the immunoreactivity was decreasing as pressure and time increase. At 2.6 bar (30 min) only two bands (19 and 16 kDa) were still detected.

The effect of DIC treatment on chickpeas and lentil allergenic proteins were similar. Untreated chickpea showed numerous IgE-binging proteins with molecular weights between 12-82 kDa. The immunoreactive band pattern after the DIC treatment at 6 bar for 3 min also showed a marked decrease in the number and intensity of IgE-binging proteins. However, in the chickpea experiment, there were no apparent distinctions among the different pressure and time conditions applied, and some more heat-stable immunoreactive proteins were still present at the same exteme DIC conditions (6 bar, 3 min).

The most effective reduction of chickpea protein antigenicity was obtained by sequential treatment with Alcalase and Flavourzyme (Clemente, Vioque, Sanchez-Vioque, Pedroche, and Millan 1999). Individual hydrolysis of chickpeas with Flavourzyme or Alcalase reached a 27% of DH; however, Alcalase hydrolysates did not show visible electrophoretic bands in contrast to those obtained with Flavourzyme.

10. Effect of Processing on Soybean (*Glycine max*) Allergens

Soybean and peanut are the two main legumes involved in hypersensitive response in numerous countries. Soybean, with around 21 known allergenic proteins, is widely consumed throughout the world. It is mainly used as ingredient in formulated foods, meat/poultry products, bakery and pastry products, dairy products, apart of its pharmaceutical and industrial applications (Endres 2001). The high consumption of soybean has also been associated with high risk of allergy for the consumers and the prevalence rate in the general population is around 0.3–0.4%. Several major allergens in soybean have been identified, namely P34 (Gly m Bd 30 K), Gly m 1, Gly m 2, Gly m 3, Gly m 4 and Gly m Bd 28 K (Verma et al. 2013). Due to the commercial use of soybean protein based food products, which frequently involves thermally processed proteins, it is of high importance to investigate the effect of heat treatment on the main allergens of this legume. Burks et al. (1991) did not find any relevant decrease of IgE-binding after heating soy proteins at various temperatures and time intervals. (Wilson, Blaschek, and Gonzalez de Mejia 2005) concluded that several procedures are needed to eliminate soyabean allergenicity, particularly P34, the major allergenic protein.

For evaluating DIC-treatment effect, untreated soybean was used as control (Cuadrado et al. 2011). All the major allergens and other minor immunoreactive proteins (13-119 kDa) were detected in the control soybean. Although DIC treatment at 3 bar for 1 and 3 min resulted in a slight reduction of the soybean immunoreactive bands, however, when pressure was increased up to 6 bar and applied for 3 min, from the immunoblot pattern any immunoreactive proteins could be detected (Cuadrado et al. 2011). Similar results were found when autoclaving was applied (2.6 bar, 30 min) on this legume (Cuadrado et al. 2007). The DIC technique employed has the strongest effect on the immunoreactive proteins of soybean compared to the other legumes studied. The short processing time (3 min) represent an advantage for future potential applications in the food industry. Elimination of allergenic proteins via processing could eventually enhance the safe use of soybean products, making them available to soy-sensitized individuals.

CONCLUSION

This review provides an update on the impact of various processing methods on legume allergies. Though few processing methods seem promising in reducing the IgE reactivity, it is still uncertain to what extent these results are clinically relevant. Various reports have indicated an increase in the cases of food allergies around the world, especially sensitivity to peanuts. The only solution for the allergic patients is complete abstinence from that particular food or ingredient which is not viable in all instances. Under the given circumstances, it is critical to understand the structural changes in the protein due to processing to comprehend whether a specific technique (conventional and novel) is aggravating or mitigating the allergenicity. Moreover, it has been demonstrated that the degree of processing can dramatically affect the results of digestibility, solubility and other parameters related with IgE-reactivity. In some allergenic proteins, the epitopes are destroyed, but they are unaltered in others. In this way, processing can alter the overall IgE binding profiles of legume proteins. However, as *in vitro* IgE-reactivity is an indicator of potential in vivo allergenicity, further *in vitro* assays and in vivo clinical data are required to confirm that these treatments can reduce the *in vivo* allergenicity of these legumes. Only after such studies, these putative hypoallergenic foods could be safely consumed and even utilised as desensitizing food.

References

Akkerdaas, J, EI Finkina, SV Balandin, S Santos Magadan, A Knulst, M Fernandez-Rivas, R Asero, R van Ree, and TV Ovchinnikova. 2012. "Lentil (Lens culinaris) Lipid Transfer Protein Len c 3: A Novel Legume Allergen." *International Archives of Allergy and Immunology* 157 (1):51-57.

Álvarez-Álvarez, J., E. Guillamón, J. F. Crespo, C. Cuadrado, C. Burbano, J. Rodríguez, C. Fernández, and M. Muzquiz. 2005. "Effects of extrusion, boiling, autoclaving, and microwave heating on lupine allergenicity." *Journal of Agricultural and Food Chemistry* 53 (4):1294-8.

Barre, A, G Jacquet, C Sordet, R Culerrier, and P Rouge. 2007. "Homology modelling and conformational analysis of IgE-binding epitopes of Ara h 3 and other legumin allergens with a cupin fold from tree nuts." *Molecular Immunology* 44 (12):3243-3255.

Besler, M, H Steinhart, and A Paschke. 2001. "Stability of food allergens and allergenicity of processed foods." *J. Chromatogr. B Biomed. Sci. Appl* 756.207-228.

Beyer, K. B., E. Morrow, Li X-M., and L. Bardina. 2001. "Effects of cooking methods on peanut allergenicity." *J. Allergy Clin.Immunol* 107:1077-1081.

Burks, A. W., L. W. Williams, W. Thresher, C. Connaughton, G. Cockrell, and R. M. Helm. 1992. "Allergenicity of peanut and soybean extracts altered by chemical or thermal denaturation in patients with atopic dermatitis and positive food challenges." *J Allergy Clin Immunol* 90:889-97.

Burks, AW, HA Sampson, and GA Bannon. 1998. "Peanut allergens." *Allergy* 53:725-730.

Burks, AW, LW Williams, R. M. Helm, W Thresher, JR Brooks, and HA Sampson. 1991. "Identification of soy protein allergens in patients with atopic dermatitis and positive soy challenges: determination of change in allergenicity after heating or enzyme digestion" *Advances in experimental medicine and biology* 289:295-307.

Cabanillas, B., M. M. Pedrosa, J. Rodriguez, M. Muzquiz, S. J. Maleki, C. Cuadrado, C. Burbano, and J. F. Crespo. 2012. "Influence of Enzymatic Hydrolysis on the Allergenicity of Roasted Peanut Protein Extract." *Int Arch Allergy Immunol* 157 (1):41-50.

Cabanillas, B, SJ. Maleki, J Rodriguez, C Burbano, M Muzquiz, MA Jimenez, MM. Pedrosa, C Cuadrado, and JF. Crespo. 2012. "Heat and pressure treatments effects on peanut allergenicity." *Food Chemistry* 132 (1):360-6.

Cabanillas, B, MM. Pedrosa, J Rodriguez, A Gonzalez, M Muzquiz, C Cuadrado, JF. Crespo, and C Burbano. 2010. "Effects of enzymatic hydrolysis on lentil allergenicity." *Mol Nutr Food Res* 54:1266-72.

Clemente, A. 2000. "Enzymatic protein hydrolysates in human nutrition." *Trends Food Sci. Technol.* 11:254-262.

Clemente, A., J. Vioque, R. Sanchez-Vioque, J. Pedroche, J. Bautista, and F. Millan. 1999. "Protein quality of chickpea cicer arietinum L., protein hydrolysates." *Food Chemistry* 67:269-274.

Clemente, A., J. Vioque, R. Sanchez-Vioque, J. Pedroche, and F. Millan. 1999. "Production of extensive chickpea Cicer arietinum L., protein hydrolysates with reduced antigenic activity." *J. Agric. Food Chem.* 47:3776-3781.

Crespo, JF, CY Pascual, AW Burks, RM Helm, and M Martin-Esteban. 1995. "Frecuency of food allergens in a pediatric population fron Spain." *Pediatr. Allergy Immunol.* 6:39-43.

Crespo, JF, and J Rodriguez. 2003. "Food allergy in adulthood." *Allergy* 58:98-113.

Cuadrado, C, B Cabanillas, MM Pedrosa, M Muzquiz, J Haddad, K Allaf, J Rodriguez, JF Crespo, and C Burbano. 2011. "Effect of Instant Controlled Pressure Drop on IgE Antibody Reactivity to Peanut,

Lentil, Chickpea and Soybean Proteins." *International Archives of Allergy and Immunology* 156 (4):397-404.

Cuadrado, C, B Cabanillas, MM Pedrosa, A Varela, E Guillamón, M Muzquiz, JF Crespo, and J Rodriguez. 2007. "Effect of heat treatments on legume allergenicity." Integrating Legume Biology for Sustainable Agriculture. Proceedings of the 6th European Conference on Grain Legumes, Lisbon, Portugal, 12-16 November.

Cuadrado, C, B Cabanillas, MM. Pedrosa, A Varela, E Guillamon, M Muzquiz, JF. Crespo, J Rodriguez, and C Burbano. 2009. "Influence of thermal processing on IgE reactivity to lentil and chickpea proteins." *Molecular Nutrition & Food Research* 53 (11):1462-1468.

Chung, S. Y., and E. T. Champagne. 1999. "Allergenicity of Maillard reaction products from peanut proteins." *J Agric Food Chem* 47:5227-5231.

Chung, Si-Yin, and S Reed. 2011. "Removing peanut allergens by tannic acid." *Food Chemistry* 134:1468-1473.

Chung, SY, and ET Champagne. 2001. "Association of end-product adducts with increased IgE binding of roasted peanuts." *J Agric Food Chem* 49:3911-3916.

Chung, SY, CL Butts, SJ Maleki, and ET Champagne. 2003. "Linking peanut allergenicity to the processes of maturation, curing, and roasting." *J Agric Food Chem* 51:4273-4277.

Davis, PJ, and SC Williams. 1998. "Protein modification by thermal processing." *Allergy* 53:102-105.

Duranti, M, F Sessa, A Scarafoni, T Bellini, and F Dallocchio. 2000. "Thermal stabilities of lupin seed conglutin ç protomers and tetramers." *J Agric Food Chem 48*:1118-1123.

Duranti, M. 2006. "Grain legume proteins and nutraceutical properties." *Fitoterapia* 77:67-82.

Endres, JG ed. 2001. *Soy protein products-characteristics, nutritional aspects, and utilization.* Champaign: AOCS Publication.

Guillamón, E, J Rodríguez, C Burbano, M Muzquiz, MM Pedrosa, B Cabanillas, JF Crespo, AI Sancho, ENC Mills, and C. Cuadrado. 2010.

"Characterization of lupin major allergens (*Lupinus albus* L)." *Mol Nutr Food Res* 54:1168-1676.

Guillamón, E., C. Burbano, C. Cuadrado, M. Muzquiz, M. M. Pedrosa, M. Sanchez, B. Cabanillas, J. F. Crespo, J. Rodriguez, J. Haddad, and K. Allaf. 2008. "Effect of an instantaneous controlled pressure drop on in vitro allergenicity to lupins (Lupinus albus var Multolupa)." *Int Arch Allergy Immunol* 145 (1):9-14.

Guillamón, E., C. Cuadrado, M. M. Pedrosa, A. Varela, B. Cabellos, M. Muzquiz, and C. Burbano. 2010. "Breadmaking properties of wheat flour supplemented with thermally processed hypoallergenic lupine flour." *Spanish Journal of Agricultural Research* 8 (1):100-108.

Guillamon, Eva, Carmen Burbano, Carmen Cuadrado, Mercedes Muzquiz, Mercedes M. Pedrosa, Monica Sanchez, Beatriz Cabanillas, Jesus F. Crespo, Julia Rodriguez, Joshep Haddad, and Karin Allaf. 2008. "Effect of an instantaneous controlled pressure drop on in vitro allergenicity to lupins (Lupinus albus var Multolupa)." *International Archives of Allergy and Immunology* 145 (1):9-14.

Humiski, L. M., and R. E. Aluko. 2007. "Physicochemicaland bitterness properties of enzymatic pea protein hydrolysates." *J Food Sci* 72:S605-S611.

Lee, H. W., E. H. Keum, S. J. Lee, D. E. Sung, D. H. Chung, S. I. Lee, and S. Oh. 2007. "Allergenicity of proteolytic hydrolysates of the soybean 11S globulin." *J Food Sci* 72 (3):C168-72.

Leterme, P. 2002. "Recommendations by health organizations for pulse consumption." *Br J Nutr* 88:239-242.

Lopez-Torrejon, G., G. Salcedo, M. Martin-Esteban, A. Diaz-Perales, C. Y. Pascual, and R. Sanchez-Monge. 2003. "Len c 1, a major allergen and vicilin from lentil seeds: Protein isolation and cDNA cloning." *Journal of Allergy and Clinical Immunology* 112 (6):1208-1215.

Lucas, J. S., and R. G. Atkinson. 2008. "What is a food allergen?." *Clin. Exp. Allergy* 38:1095-1099.

Maleki, S. J., S. Y. Chung, E. T. Champagne, and J. P. Raufman. 2000a. "The effects of roasting on the allergenic properties of peanut proteins." *Journal of Allergy and Clinical Immunology* 106 (4):763-768.

Maleki, S. J., S. Y. Chung, E. T. Champagne, and J. P. Raufman. 2000b. "The effects of roasting on the allergenic properties of peanut proteins." *J Allergy Clin Immunol* 106 (4):763-8.

Maleki, S. J., O. Viquez, T. Jacks, H. Dodo, E. T. Champagne, S. Y. Chung, and S. J. Landry. 2003. "The major peanut allergen, Ara h 2, functions as a trypsin inhibitor, and roasting enhances this function." *Journal of Allergy and Clinical Immunology* 112 (1):190-195.

Maleki, Soheila J. 2004. "Food processing: effects on allergenicity." *Current opinion in allergy and clinical immunology* 4 (3):241-5.

Merritt, R. J., M. Carter, M. Haight, and L. D. Eisenberg. 1990. "Whey protein hydrolysate formula for infants with gastrointestinal intolerance to cow milk and soy protein in infant formulas." *J. Pediatr. Gastroenterol. Nutr.* 11:78-82.

Mills, C, AR Mackie, P Burney, KB Beyer, L Frewer, C Madsen, E Botjes, RW Crevel, and R van Ree. 2007. "The prevalence, cost and basis of food allergy across Europe." *Allergy* 62 (7):717-722.

Mills, ENC, J A Jenkins, MJC Alcocer, and PR Shewry. 2004. "Structural, biological, and evolutionary relationships of plant food allergens sensitizing via the gastrointestinal tract." *Critical reviews in food and nutrition* 44:379-407.

Mills, ENC, J Jenkins, N Marigheto, PS Belton, AP Gunning, and VJ Morris. 2002. "Allergens of the cupin superfamily." *Biochemical Society Transactions* 30 (6):925-929.

Mondoulet, L., E. Paty, M.-F. Drumare, and S. et al. Ah-Leung. 2005. "Influence of thermal processing on the allergenicity of peanut proteins." *J. Agric. Food Chem 53*:4547-4553.

Moneret-Vautrin, DA, L Guérin, G Kanny, J Flabbee, S Frémont, and M Morisset. 1999. "Cross-allergenicity of peanut and lupine: the risk of lupine allergy in patients allergic to peanuts." *Journal of Allergy Clinical Immunology* 04 (4 Pt 1):883-888.

Mueller, G A, R A Gosavi, A Pomes, S Wuenschmann, A F Moon, R E London, and L C Pedersen. 2011. "Ara h 2: crystal structure and IgE binding distinguish two subpopulations of peanut allergic patients by epitope diversity." *Allergy* 66 (7):878-885.

Niphadkar, P. V., S. P. Patil, and M. M. Bapat. 1997. "Chickpea induced anaphylaxis." *Allergy* 52:480-482.

Patil, S. P., P. V. Niphadkar, and M. M. Bapat. 2001. "Chickpea: a major food allergen in the Indian subcontinent and its clinical and immunochemical correlation." *Annals of Allergy Asthma & Immunology* 87 (2):140-145.

Rabjohn, P, EM Helm, JS Stanley, CM West, HA Sampson, AW Burks, and GA Bannon. 1999. "Molecular cloning and epitope analysis of the peanut allergen Ara h 3." *J Clin Invest* 103:535-542.

Sanchez-Monge, R., C Y Pascual, A. Diaz-Perales, J Fernandez-Crespo, M Martin-Esteban, and G Salcedo. 2000. "Isolation and characterization of relevant allergens from boiled lentils." *Journal of Allergy and Clinical Immunology* 106 (5):955-961.

Sathe, S K., and G M Sharma. 2009. "Effects of food processing on food allergens." *Molecular Nutrition & Food Research* 53 (8):970-978.

Schmitt, DA, JB Nesbit, BK Hurlburt, H Cheng, and SJ Maleki. 2010. "Processing Can Alter the Properties of Peanut Extract Preparations." *Journal of Agricultural and Food Chemistry* 58 (2):1138-1143.

Seo, W. H., H. G. Lee, and H. H. Baek. 2008. "Evaluation of bitterness in enzymatic hydrolysates of soy protein isolate by taste dilution analysis." *J. Food Sci.* 73: S41-S46.

Sicherer, SH, and HA Sampson. 2007. "Peanut allergy: emerging concepts and approaches for an apparent epidemic." *Journal of Allergy and Clinical Immunology* 120 (3):491-503.

Sormus de Castro Pinto, S. E., V. A. Neves, and B. M. Machado de Medeiros. 2009. "Enzymatic hydrolysis of sweet lupin, chickpea, and lentil 11S globulins decreases their antigenic activity." *J Agric Food Chem* 57 (3):1070-5.

Terracciano, L., P. Isoardi, S. Arrigoni, A. Zoja, and A. Martelli. 2002. "Use of hydrolysates in the treatment of cows milk allergy." *Ann. Allergy Asthma Immunol.* 89:86-90.

Verma, A. K., S. Kumar, M. Das, and P. D. Dwivedi. 2013. "A Comprehensive Review of Legume Allergy." *Clin Rev Allergy Immunol* 45:36-46.

Vioque, J., R. Sanchez-Vioque, A. Clemente, J. Pedroche, J. Bautista, and F. Millan. 1999. "Purification and partial characterization of chickpea 2S albumin." *Journal of Agricultural and Food Chemistry* 47 (4):1405-1409.

Wal, JM. 2003. "Thermal processing and allergenicity of foods." *Allergy* 58:727-729.

Wickham, M., R. Faulks, and C. Mills. 2009. "In vitro digestion methods for assessing the effect of food structure on allergen breakdown." *Mol Nutr Food Res* 53 (8):952-8.

Wilson, S, K Blaschek, and E Gonzalez de Mejia. 2005. "Allergenic proteins in soybean: processing and reduction of P34 allergenicity." *Nutr Rev* 63:47-58.

In: Legumes for Global Food Security
Editors: Alfonso Clemente et al.
ISBN: 978-1-53612-265-7

Chapter 10

THE FUNCTION OF CONGLUTIN PROTEINS FROM THE "SWEET LUPIN" GROUP IN FOOD ALLERGIES

Elena Lima-Cabello[1], Jose M. Berral-Hens[1], Jose F. Florido[2], Juan D. Alché[1] and Jose C. Jimenez-Lopez[1,*1]

[1]Department of Biochemistry,
Cell and Molecular Biology of Plants,
Estacion Experimental del Zaidin,
Spanish National Research Council (CSIC), Granada, Spain
[2]Allergy Service, "San Cecilio" University Hospital, Granada, Spain

ABSTRACT

During the last fifteen years, functional foods have gained attention from the general population, and from the food industry because of their implications in upholding of the health and the wellbeing.

[1] Corresponding Author Email: josecarlos.jimenez@eez.csic.es.

Sweet lupin is the common name that represents four domestic species, *Lupinus albus* (white lupin), *Lupinus angustifolius* (narrow-leafed lupin), *Lupinus luteus* (yellow lupin), and *Lupinus mutabilis* (pearl lupin). Sweet lupin is increasingly used in food, especially in European countries and Australia since the last few years, and particularly in the bakery industry. Seed from this group of lupin species can be consider as functional food, since multiple health benefits have been demonstrated i.e., prevention of cardiovascular disease, dyslipidemia, and glycaemia control.

However, as lupin is becoming more popular in food preparation, the number of cases of allergy has increased among the population. Lupin allergy seems to arise as both primary sensitization and as result of cross-reactivity to other legumes, particularly peanut. Although lupin is implicated to cause allergy, clinical reactivity to lupin flour may be less relevant when compared to sensitization. Furthermore, when IgE cross-reactivity between proteins of a legume family occurs, it does not necessarily translate to clinically symptomatic allergies. Additionally, clinical hypersensitivity to one legume does not warrant dietary elimination of all legumes.

This work aims to provide information of lupin allergy, in order to help understanding different aspects, i.e., primary lupin allergy sensitization, cross-allergenicity to other legumes such as peanut, prevalence, clinical relevance, allergen proteins characterization, and to minimize risks for those whom may suspect of developing any atopic reaction. The key to prevent lupin allergy is to adopt measures with concerted efforts from the various sectors like policy makers, lupin manufacturers, healthcare professionals and the general consumer.

Keywords: allergen proteins, cross-reactivity, IgE-binding, *L. angustifolius*, *L. albus*, *L. luteus*, *L. mutabilis*, legume, PULSE, prevalence, sensitization, vicilin

1. INTRODUCTION

The genus *Lupinus* belongs to the *Leguminosae* family, which amongst others also contains PULSEs such peanut and other legumes as soybean and many others (http://www.pulsecanada.com/about-us/what-is-a-pulse). There are more than 450 *Lupinus* species ranging from wild species to these cultivated. Four of these species called "sweet lupin" have importance for the agriculture and food fields: *Lupinus albus* in the Mediterranean area, *Lupinus angustifolius* in Australia, *Lupinus luteus* in Central Europe, and *Lupinus mutabilis* in South America.

Lupin is a useful spring-sown green manure crop, especially on light soils. It is cultivated globally for animal feed, but also to be ploughed under as a soil enhancer (Peeters et al. 2007). Most of the species from *Lupinus* genus contain alkaloids that result toxic for human. Only sweet lupin group contains low levels (0.05%) of these toxic chemicals making them suitable for human consumption.

Lupin is increasingly used as a protein source in European countries where lupin is used to replace genetic modified soy derived products (Campbel et al. 2007; Peeters et al. 2009), and to substitute animal proteins, i.e., milk and egg white, since lupin flour is an excellent source of proteins, having a high content of proteins as albumin (55.6%) than globulins (31.5%).

Lupin was officially introduced as a food ingredient, and the inclusion of lupin in wheat flour was first permitted in the United Kingdom in 1996, in France at the end of 1997 and in Australia in 2001 to add protein and fibre and to improve food texture (Faeste, 2010). Since then, it has been increasingly used as a functional food or food ingredient because of its nutritional properties such as high content of protein, fiber, oligosaccharides and phenolic compounds, as well as its satisfactory technological properties, i.e., bakery products may contain up to 10% lupin flour (Duranti et al. 2006).

Interestingly, lupin seed components have been associated with the prevention of obesity, dyslipidemia, eventually cardiovascular disease, and diabetes (Arnoldi et al. 2015). Lupin is also recommended for patients with wheat protein allergy and coeliac disease (Rotiroti et al. 2007). Recently, it has been demonstrated *in vitro* interaction of γ-conglutin, and more recently β-conglutin proteins to insulin (Terruzzi et al. 2011; Lima-Cabello et al. 2017). Gamma-conglutin decrease the glycaemia in animals of experimentation (Terruzzi et al. 2011), and β-conglutin proteins induce a modulation of the expression levels of key genes of the insulin pathway as prevention and potential treatment of type 2 diabetes, as well as to the regulation of pro-inflammatory genes for inflammatory diseases amelioration (Lima-Cabello et al. 2017).

Since the first case of occupational allergy to lupin (Parisot et al. 2001), the number of reports on allergic reactions to lupin flour, seed, or dust is rising in number. During the following years allergic reactions also after respiratory exposure mostly under occupational conditions have been described (Crespo et al. 2001; Campbell et al. 2007). The number of cases of lupin allergy and the severity of the symptoms indicate the importance of this new up-coming allergy and the relevance of these allergenic proteins.

2. Allergenic Proteins in the Lupin Seed

One of the most abundant lupin seed storage proteins are globulins, integrated by two major protein types: β-conglutin (7S globulin or vicilin-like protein) and α-conglutin (11S globulin or legumin-like proteins). Also, two minor components: δ-conglutin (2S sulfur-rich albumin) and γ-conglutin (basic 7S globulin (Duranti et al. 2008) are also relevant components. Beta-conglutins show the greatest structural variation (Lima-Cabello et al. 2016), being composed of over 20 polypeptide chains with no disulfide bridges, covering a broad range of molecular masses (MW 10–80 kDa). Alpha-conglutins are composed of a heterogeneous mixture of subunits linked by two disulfide bridges (acid subunits of 54–47 kDa and basic subunit of 20 kDa). Gamma-conglutin consists of disulfide-linked of 30 and 17 kDa polypeptides, and delta-conglutin, are made up of two disulphide-linked polypeptide chains of 9 and 4 kDa (Duranti et al. 2008).

Among seed proteins, the following have been identified as allergens such as β-conglutin, α-conglutin (Dooper et al. 2006), γ-conglutin (Magni et al. 2005, 2007), Lup a LTP (a lipid transfer protein) (Asero et al. 2007), and a 2S albumin (Magni et al. 2005, 2007; Wait et al. 2005), whose isolated fraction appears to consist of a number of isoforms ranging from 4 to 11 kDa. Nevertheless, the lupin conglutins α and γ are proposed as candidate allergens (Magni et al. 2005). Beta-conglutin has been identified as allergen of *L. angustifolius*, and designated as Lup an 1 (Goggin et al. 2008). Recently, a comparable protein from *L. albus* has been identified as allergen (Lup a 1) and incorporated as a new lupin allergen (https://www.allergome.org) belonging to β-conglutin family as well. Lup an 1 has a sequence similar to the peanut vicilin, Ara h 1, which has been most commonly implicated for allergenicity (Sanz et al. 2010; Ballabio et al. 2013).

Most of the polypeptides that integrate the different families of lupin seed proteins (particularly the β-conglutin family involved in food allergy), have been recognized by their MW and characterized serologically (IgE binding capacity). Multiple IgE-binding proteins with molecular weights ranging between 10 and 80 kDa have been identified by immunoblotting in both patients with allergy to lupin and other legumes and patients with lupin allergy alone throughout numerous studies, which are listed next (Prieto et al. 2010):

- The use of 23 patient sera with lupin-specific IgEs allowed identifying two major allergens from *L. albus* with MW of 34.5 (Lup-1) and 20 kDa (Lup-2), and a 43 kDa protein displaying IgE cross-reactivity with peanut (Guillamón et al. 2010).
- Lupin allergenic proteins of 71, 59 and 34 kDa in size, along with other 24 and 17 kDa proteins were isolated (Parisot et al. 2001). In addition, sensitization to 9 and 10 kDa proteins was described in raw and cooked seeds (Rosales et al. 2002).
- Heat-stable allergens in the range of 35 to 55 kDa were isolated, with a 21 kDa major polypeptide (Hefle et al. 1994).
- The use of sera from 6 atopic patients to lupin flour helped to discriminate between lupin and peanut allergen sources. Thus, two of them (non-allergic to peanut) displayed IgE binding to 66 kDa proteins, and weak binding to 14 and 24 kDa proteins, whereas patients with both peanut and lupin allergy showed weak binding to Lupin proteins of about 66 kDa or 14 to 21kDa (Peeters et al. 2007).

- The sera of 2 patients with oral allergic syndrome (characterized by lupin ingestion) recognized proteins of 34 kDa and 40-65 kDa. Serum of only one of them also recognized 7-15, 38, 77, 162 and 195 kDa proteins (Quaresma et al. 2007).
- A 14 kDa polypeptide was identified in an investigation of lupin allergen in biscuits, chicken bouillon and a dehydrated chicken soup (Rojas-Hijazo et al. 2006).
- Moneret and colleagues (Moneret-Vautrin et al. 1999) identified a 43-kDa IgE-binding protein, possibly corresponding to γ-conglutin, in a patient with known peanut allergy.
- Sanz and colleagues (Sanz et al. 2010) described a 29-kDa band in a patient with isolated lupin allergy.
- Wüthrich (Wüthrich 2008) identified a 50-kDa band not inhibited by peanut extract in a patient with lupin allergy who tolerated other legumes.
- Goggin and colleagues (Goggin et al. 2008) identified ß-conglutin, designated as Lup an 1, as a major allergen of lupin.

Multiple polypeptides from a large MW range have been identified with IgE-binding capacity, particularly in *L. angustifolius* and *L. albus*. These peptides may possess ability to trigger allergy symptoms among the population reacting to a potential hidden food allergen preceded by former sensitization via inhalation. Further studies are needed to determine which lupin allergens might predict tolerance to other legumes and which might be related to legume polysensitization.

3. Lupin Allergy Prevalence

The frequency of lupin-allergic reaction in the general population is not well-known. Such prevalence may depend on the daily food intake as well as the geographical localization (Crespo et al. 1995; Verma et al. 2013). What seems to be clear is that the prevalence of lupin allergy increases in parallel with consumption of this plant. The increasingly use of lupin flour in food preparation is a big concern, particularly because lupin is often present as a hidden allergen in processed foods (Hefle et al. 1994; Smith et al. 2004; Faeste et al. 2004). Lupin is ninth in the rank order of "risk" allergens. On the other hand, the reactivity threshold for this allergen is low: 250 to 1000 mg are enough to initiate the symptoms (Peeters et al. 2009). These doses correspond to amounts found in routinely consumed lupin-containing products.

It has been found three possible forms of clinical allergy to lupin:

1. Primary allergy to peanut with cross-reactivity to lupin;
2. Primary allergy to lupin;
3. Primary allergy by inhalation of lupin pollen or flour (often occupational).

The origin of sensitization remains unknown. It may be either routine ingestion or primary inhalation of lupin pollen followed by cross sensitization to lupin seeds. Cross reactivity between pollen and seed has been demonstrated *in vitro* (Moneret-Vautrin et al. 1999). Another possibility is occupational sensitization due to inhalation of lupin flour (Parisot et al. 2001; Crespo et al. 2001; Campbell et al. 2007).

The prevalence of allergic reactions to lupin flour or seed appears to be increasing along with the practice of adding protein-rich flour to bread. These flour and seed are known to commonly be allergic also to other legumes, particularly peanut, soya and pea. There is a relatively high frequency of primary sensitization to lupin flour, as confirmed by the fact that a study of atopic children exposed a 8% of sensitization rate, 25% with primary sensitization and 75% with cross sensitization to peanut (Mazeyrat et al. 2004).

Legume sensitization may be a dynamic process, as described in a patient whose symptoms started with reactions to only a single legume and who then progressively developed sensitization to lentil, white bean, lupin, and pea (Matheu et al. 1999). Alternatively, clinical hypersensitivity to one legume does not need dietary elimination of all legumes.

Nowadays, lupin allergy is predominantly observed in those patients with peanut allergy, and assessed by sensitization methods such as skin prick test and/or serum IgE positivity, as well as clinical reactivity (oral challenge test) to lupin flour (Gayraud et al. 2009). Such subjects allergic to peanuts with allergy symptoms also to lupin range 16 to 63%, being the primary sensitization to lupin between 1.8 to 3.7%.

Prevalence is also dependent of *the population age*: lupin allergy affects 5.2% of children aged up to 15 years and 1.7% aged between 15 to 30 years. It has not been reported so far to affect infants according to our knowledge.

Prevalence is also dependent of *the population gender*, fact that is well-known (Lalles et al. 1996; Pereira et al. 2002). Looking at lupin atopic adults, there is a female predominance (62.4%), whilst for peanut allergy there was only slight majority (52%). The frequency of peanut allergy among children was 11.2% (Gayraud et al. 2009).

The methodology to detect sensitization is also important: prick-tests detection of peanut allergy compared to simple non-allergic sensitization indicated that 72% of sensitized children were allergic, but only 25% of sensitized adults. This may indicate that tolerance mechanisms are more effective in adults. 17% of children and 14.5% of adults allergic to peanut showed cross sensitization to lupin. This cross sensitization could be different in another study. Thus, the ratio sensitization/allergy will have to be evaluated in further studies including double-blind, placebo-controlled food challenge (DBPCFC) (Shaw et al. 2008; Lindvik et al. 2008)

Geographic location is another factor to be take into account, since the prevalence of lupin allergy has increased markedly in some countries, especially France, where the addition of lupin flour to wheat flour was first permitted in 1997 (Moneret-Vautrin et al. 1999). The first report of lupin anaphylaxis was in 1999 (Smith et al. 2004). In 2002, lupin was the fourth most frequent cause of severe food-associated anaphylaxis reported to the French Allergy Vigilance Network. In a lupin sensitization study in Portugal among 1,160 subjects used that were positive in Skin Prick Test (SPT) to Lupin, lupin sensitization rate was of 4.1%, whereas 82.1% co-sensitization between lupin and pollen, 75% co-sensitization between lupin and legumes, and 28.5% co-sensitization between lupin and latex were also detected (Reis et al. 2007).

4. Lupin Allergy: Primary Sensitization

There are clear evidences that occupational allergy to airborne lupin flour must be considered as one of the main sources of primary sensitization to lupin (Quaresma et al. 2007).

However, this primary sensitization may also be associated with sensitization to other legumes (mainly peanut) (Magni et al. 2005). As example of lupin primary sensitization there are recent reports showing that lupin allergy occurs for the first time in people who are not allergic to peanut or legumes. In 2014, lupin allergy was reported taking place for the first time in three adults, where none had any previous history of peanut, legume or any other food allergy (Bansal et al. 2014).

The methodologies to be used to assess sensitization could be the combination of the clinical picture, skin prick tests and serum IgE to lupin and to peanut for sensitization assessment to lupin not related to peanut allergy. To finalize the assessment, the complete inhibition of lupin-specific IgE by lupin extract (90%) vs the only partial inhibition by peanut extract (65% and 75%) strongly supports the initial results. Studies using immunoblotting performed using IEF as well as SDS PAGE showed that some lupin IgE-binding proteins were inhibited by peanut extract, but there are also bands only partially inhibited and others no inhibited at all. These results clearly indicate that lupin is capable of inducing allergic sensitization out of the context of cross-reactivity with peanut (Quaresma et al. 2007).

5. Lupin Allergy: Cross-Reactivity

Detection of allergen-specific IgE antibodies simply identifies sensitization to a particular allergen(s) and does not allow a decisive differentiation between clinically relevant IgE reactivity and IgE reactivity not accompanied by clinical symptoms (Crespo et al. 2004).

Otherwise, IgE-positivity to several legumes basically indicates an immunological cross-reactivity via linear epitopes (sequence homology) or similar conformational epitopes without allowing estimation or prognosis of the clinical significance. The following legumes may cross-react with lupin as published by several groups: peanut, soybean, lentils, beans, chickpeas, and peas.

Numerous different vicilin-like proteins have been identified as major allergens of some legumes, including Ara h 1 (*A. hypogaea*), Len c 1 (*L. culinaris*), Gly m Bd 28K y 60K (*G. max*), Gly m 5 and 6 (Bublin et al. 2014). The high sequence homology among vicilin proteins from legumes (peanut, pea, lentil, and soybean) seems to be responsible for the IgE cross-reactivity detected between these grain legumes (Guillamón et al. 2010). In a similar fashion, vicilin allergens such as Ara h 1 and Lup-1 also share high homology (Chruszcz et al. 2011).

The immuno-cross-reactivity between lupin and other legumes has been previously demonstrated, particularly in lupin, peanut and pea (Bublin et al. 2014). However, such IgE reactivity may not always be related to clinical cross-reactivity. This serological cross-reactivity may be due to the high sequence homology found among some of their allergens (Guillamón et al. 2008).

Despite the high homology among these different proteins from individual species of the *Fabaceae* (legume family), extensive cross-reactivity could be expected but in fact does not occur frequently (Yman et al. 1982). In an *in vitro* study, the IgE binding to protein extracts of 11 food legumes was examined by IgE antibody determination and RAST inhibition. Cross-allergenicity was demonstrated to be most marked among the extracts of peanut, garden pea, chick pea, and soybean (Barnett et al. 1987). However, clinical studies have found little cross-reactivity among members of the legume family, with the possible exception of cross-reactivity between lupin seed and peanut (Bernhisel et al. 1989; Eigenmann et al. 1996).

Therefore, the high identity between Lup a 1 and legume IgE-binding vicilins from lentil, peanut, soybean and pea favor IgE-binding using *L. albus* protein extracts. Lup a 1 was identified as the lupin β-conglutin precursor and was highly homologous to soy β-conglycinin. This fact is consistent with the findings of Magni et al. (Magni et al. 2012) who observed that the 20-kDa band (Lup a 2), which strongly reacted with serum IgE from lupin sensitized patients, corresponded to the basic subunit of α-conglutin (lupin legumin-like protein).

Serological cross-reactivity between members of the legume family does not always result in clinically symptomatic allergies (Islam et al. 2012). In a study carried out in France and Belgium, cross-reactivity with lupin was found to be more frequent in those with peanut allergy (17.1% and 14.6%, respectively) (Gayraud et al. 2009). Cross-allergic clinical reactions to members of the legume family such as soybean and lentils occur in about 5% of peanut-allergic patients, but cross-allergic reactions were found to be 68% between peanut and Lupin (Bernhisel et al. 1989). Peanut-Lupin allergy is high, contrary to the risk with other legumes, since 68% of a group of peanut-allergic patients showed positive reactions to lupin flour when tested (Bahr et al. 2014). Subjects with positive SPT to lupin also reported a history of reactions to green pea (Hefle et al. 1994).

There is cross-reactivity between peanut and other legumes, i.e., soybean, pea, bean, and lentils in nearly 5% of the peanut-allergic patients (Bernhisel-Broadbent et al. 1989). According to Moneret-Vautrin et al. (1999) the risk of peanut-allergic patients to react to lupin with allergic symptoms is 28%.

An investigation from UK with peanut allergic children and adolescents revealed a sensitization rate to lupin of 34% and an estimated prevalence of lupin allergy of 8% (Bahr et al. 2014).

Mazeyrat et al. (2004) study showed that 5% of atopic subjects were sensitized to lupin; from these, 31% with lupin sensitization were allergic to other legumes (peas, lentils, beans). Peanut allergen-specific IgE-antibodies were found in higher concentrations in subject with peanut plus lupin sensitization compared with subjects exclusively sensitized to peanut. These findings demonstrated that reactivity against legumes like peas, lentils, or beans is an important risk factor for the sensitization against lupin.

Reis and colleagues investigations found lupin-specific IgE-antibodies and detected a sensitization rate of 4.1%; 75% a co-sensitization between lupin and other legumes, and 82.1% a co-sensitization between lupin and pollen (Reis et al. 2007).

Peeters et al. (2009) showed that 82% of the study population was sensitized to lupin, 55% to pea, and 87% to soy. Clinically relevant sensitization to lupin, pea, or soy occurred in 35, 29, and 33%, respectively, of the study population. None of the patients had been aware of the use of lupin in food (Peeters et al. 2009).

In a study of severe anaphylaxis to almond, a 45 and a 30 kDa protein isolated from almond showed a 60% homology to the γ-conglutin protein from lupin seed and to the 7S globulin from soybean respectively. Developing immunoblot inhibition experiments were detected the IgE binding to almond 2S albumin and γ-conglutin in the presence of cross-reacting walnut or hazel nut antigens (Poltronieri et al. 2002). There were similar findings in other combinations, with 2S albumin from sunflower seeds, homologous with albumin from lupin seeds (δ-conglutin) (Egorov et al. 1996).

Therefore, as the level of peanut-specific IgE was significantly correlated to the levels of lupin-, pea- and soy-specific IgE, similar IgE binding properties of these allergens (cross-reactivity) could be suggested. However, IgE-inhibition experiments are needed, to investigate and to identify allergens within lupin responsible for cross-reactivity.

Recently, it was shown that patients allergic to lupin but not to peanut displayed IgE binding to other lupin allergens when compared with the patients with a combined allergy (Peeters et al. 2009), which indicates that there are remarkable differences in allergen recognition between these two groups of patients. Having symptoms to peanut or not had no effect on the grade of sensitization to lupin and on the frequency of symptoms to lupin.

This illustrates that the peanut-sensitized patients without symptoms to peanut have comparable high risks for being lupin-allergic as the peanut-allergic patients.

6. Lupin-Containing Food Allergy Features

Lupin flour is being progressively more used in the food industry because of its high protein content and for its excellent technological characteristics. Lupin is also a valuable alternative to soybean avoiding genetically modified food ingredients. It also provides a high proportion of proteins (>35%) and fiber (25-30%), very low in lipids (8-10%), and carbohydrates (Lucas et al. 2015), as well as improved texture. Lupin does not contain gluten, thus it could be a functional ingredient in gluten-free foods (Scarafoni et al. 2009).

Lupinus albus and *L. angustifolius* are important source of proteins for human beings: these seeds have a high essential amino acid index (EAAI) and protein efficiency ratio (PER), based on lysine and tyrosine availability (Duranti et al. 2008). Lupin flour is widely considered an excellent raw material for supplementing different food products for its high protein content to be used as eggs proteins (biological value of 91%) substitute in cakes, pancakes, biscuits, or brioche or added to pasta, crisps, and bread. It has been also used as a butter substitute in cake, brioche, and croissant. The lupin content in bread varies from 0% to 10% and from 0.5% to 3% in pastry, and currently, approximately 0.11 kg lupin flour and 5.5 kg lupin flour-containing products are consumed per person per year in the Netherlands. There are other potential uses for lupin flour. For example, a German company includes lupin flour in some of its burgers and sausages. Lupin kernel fiber has also a potential as a human food ingredient as it has been used in the production of palatable fiber-enriched baked goods and pasta (Clark & Johnson, 2002).

Allergic reactions and even in some cases anaphylactic reactions have been reported since lupin introduction as food (Bansal et al. 2014). Sensitization to lupin has been reported not only to occur by consumption but also by inhalation of lupin flour by bakers (Novembre et al. 1999; Crespo et al. 2001; Moreno-Ancillo et al. 2005; Brennecke et al. 2007; Campbell et al. 2007). Some cases of primary lupin allergy have also been described (Prieto et al. 2010), but most of the allergic reactions to lupin have been reported in individuals as cross-reactivity (Bublin et al. 2014), patients with allergies to other legumes, particularly peanut, which rate around 30%. The population sensitized to peanut, which represents about 0.7–1.5% of the European population, is the most susceptible to be at risk of developing sensitization to lupin (Holden et al. 2008; Rojas-Hijazo et al. 2006; Magni et al. 2005; Dooper et al. 2007; Jappe et al. 2007; Guarneri et al. 2005).

Since 2004 European food producers are required by law to list potential allergens on packaging (gluten, fish, crustaceans, egg, soy, milk, nuts, dairy products, peanuts, mustard, celery, sesame seeds, and sulfites). On December 22, 2006, lupin was added to Appendix III A of directive 2000/13/EC (Directive 2006/142/EC). Until the provisions are fully implemented, legumes will continue to be a "hidden" allergen. However, lists of the ingredients must appear under all circumstances on the labeling of food stuffs (https://ec.europa.eu/food/safety/labelling_nutrition_en). The addition of lupin to the list of ingredients that must be labeled on prepackaged foodstuffs pursuant to EU law (https://ec.europa.eu/food/safety /labelling_nutrition/labelling_legislation_en), independently of the level, is a preventive measure aiming to protect consumers from the risks of develops allergic reactions to lupin-containing foods.

Public information is the best way to create awareness about food lupin content for these atopic patients and these with other allergies to legume, seeds and dry fruits. Information on lupin allergy and on the fact that it is an ingredient masked in other foods should be circulated more widely to the general public. Lupin is sometimes labeled as lupin, lupin flour, lupin seed or lupin bean. Under European law, lupin must always be declared and highlighted on the label when it appears in pre-packed food. Mandatory labeling, based on the analysis of published cases of allergy and anaphylaxis, is all the more justified. The food allergen labeling laws now also applies to the catering sector. When eating out or buying takeaway food, food businesses will be required to provide information on allergenic ingredients. This information can be provided in writing and/or orally.

Lupin allergens are relatively heat stable and this further complicates the problem of lupin allergy, where microwave cooking, boiling and extrusion cooking produce minimal changes on IgE binding to lupin proteins (Alvarez-Alvarez et al. 2005). Overall, many of these lupin allergens are thermostable, even in combination with pressure (Guillamón et al. 2008). However, there have been varied reports on the effects of heat on the allergenicity to legumes (Abrams and Gerstner, 2015). Particular reports showed a reduction in allergenicity after autoclaving at 138 °C for 20 minutes, particularly IgE antibodies that recognized allergenic protein bands at 23 and 29 kDa, since this method abolished the IgE binding. Thus, prolonged autoclaving might have an important effect on the allergenicity of Lupin for the majority of patients. Furthermore, a combination of heat and pressure eliminate the allergenic potency of lupin proteins: boiling in an autoclave at 121°C (1.2 bars) for 20 min significantly reduces IgE binding and at 138°C (2.6 bars) for 30 min abolished the lupin allergenicity (Verma et al. 2012). Another treatment called "instantaneous controlled pressure drop" is a novel and highly controlled process used in food technology, in which steam pressure (up to 8 bar) with heat (up to 170°C) are combined for a short time of up to 3 min. This procedure has been shown to reduce the allergenicity of lupin (Guillamón et al. 2008). On the other hand, we need to evaluate the effects of autoclaving on the solubility of proteins and effects on nutritional value of lupin.

7. Clinical Features of Lupin Allergy

From the literature reviewed above, it can be assumed that individuals with previous legume allergy have a higher risk of development of lupin sensitization, but the clinical relevance of this has not been established yet (Campbell et al. 2007). On the contrary, clinically relevant lupin allergy is often observed in individuals without a history of food allergies. Surprisingly, lupin allergy is observed more in a country where lupin is not traditionally used; implying that short-term, moderate use of lupin can result in serious allergic reactions. Asymptomatic sensitization is most frequent, but sporadically, clinical manifestations of allergy can also occur (Indoor Allergens, 1993). Lupin allergy may arise either by primary sensitization, or by clinical cross-reactivity in peanut-allergic persons (Quaresma et al. 2007; Wassenberg and Hofer, 2007). Case reports of individuals experiencing oral allergy syndrome to lupin have been known (Bansal et al. 2014).

Sensitization as well as clinical manifestations of lupin allergy can occur after seed/flour ingestion as well as pollen/flour inhalation (de Jong et al. 2010). Symptoms of lupin allergy are similar to those observed with other food or inhalant allergens and can be varied in severity (Bansal et al. 2014). The symptoms triggered after ingestion of the food or inhalation of the lupin flour may come on rapidly and include from nettle rash or urticaria, or itchy feeling in the mouth, to swelling in the face, throat and/or mouth, difficulty breathing, severe asthma, abdominal pain, nausea and vomiting. In extreme cases an anaphylactic shock leading to collapse and unconsciousness. Compared to individuals with known food allergy, individuals with primary sensitization, the allergic symptoms are severe, probably because of lack of awareness of allergy.

Lupin allergy symptoms of *primary sensitization* has been described in a person who experienced repeated episodes of rhinitis, conjunctivitis and palpebral angioedema related to her handling of lupin flour used for skin prick tests and oral challenges (Parisot et al. 2001).

Occupational allergy may occur to lupin, particularly in mill workers. It have been described that workers with allergic symptoms to lupin flour after flour inhalation. Skin reactivity to lupin seed flour extract was detected as well as lupin-specific IgE antibodies (Crespo et al. 2001).

Cases of anaphylaxis by *cross-reactivity* have been described in peanut-allergic individuals after eating a meal in a restaurant. Symptoms were pruritus of her mouth and lips, and swelling tongue as well. Few minutes later they were experiencing difficulty in breathing as their throat had "narrowed." Lupin was identified as the cause as it was present in the onion ring butter. Analysis of SPT and IgE antibodies to lupin was positive (Radcliffe et al. 2005).

There are also other cases where *cross-reactivity* has been described *between lupin and no-legume seed/grains*. A male patient reported numerous episodes of angioedema after ingestion of snacks that included lupin seeds. SPT was positive for a number of tree nuts as well as sunflower seed, as well as against lupin seed, and IgE antibodies were found to Lupin. Another individual was reported to suffer angioedema of the lip and generalized urticaria after ingestion of several lupin seeds. Skin reactivity was positive for maize flour, hazel nut, peanut, walnut, sunflower seed, chestnut, green bean, tomato, lettuce and mustard. Skin reactivity was also positive for lupin seed, and the lupin IgE antibody level was raised (Rosales et al. 2002).

The main evidence for the close relationship between peanut allergy and lupin allergy has been described and demonstrated in different studies. As example, the analysis of 24 peanut-allergic patients, where 11 of them with SPT positive to lupin flour showed positive response in 7 out 8 subject after oral challenges using the same dose as used with peanut. In summary research studies tell us that cases of isolated allergy to lupin flour, without pre-existence of peanut allergy as well as workplace asthma by inhalation, are less frequent (Leduc et al. 2002).

7.1. Allergy Eliciting Dose

There are no many studies reporting the allergy-eliciting dose of lupin, and those available found significant differences. In a study developed on eight lupin-allergic patients, adverse effect where not observed for 0.1 mg lupin flour, it was only present subjective symptoms. However, using 1 mg of lupin flour, objective symptoms were visible. Three patients experienced only subjective responses at the highest dose of 3000 mg.

The lowest eliciting dose described by Moneret-Vautrin and colleagues was 265 mg of lupin flour which induced abdominal pain and asthma in patients who were allergic to peanuts (Moneret-Vautrin et al. 1999).

Another study describes how positive clinical reactions were observed in two peanut-allergic children on consumption of 50 mg and 1.6 g of lupin proteins, respectively (Fiocchi et al. 2009).

According to the recently published VITAL (Voluntary Incidental Trace Allergen Labeling) program of The Allergen Bureau of Australia and New Zealand (ABA), the reference allergenic dose for lupin protein is 4 mg, while that for peanut was much lower (0.2 mg of peanut protein) (Remington et al. 2016).

7.2. Clinical Diagnosis of Lupin Allergy

Clinical examination and a proper history of the atopic patient is the key of diagnosis of lupin allergy. However, asymptomatic sensitization and further confirmation of allergy can be done using *in vivo* skin prick test and specific IgE measurement. Cross-reactivity can also be diagnosed by these tests. *In vitro* tests include ELISA-IgE and immunoblotting (Nicolaou and Custovic, 2011). Interestingly, it will be necessary to develop a test for lup an 1 detection and quantification in processed food and industry manufacturing in order to meet the food allergen labeling requirements of the EU, making aware of the present in food of lupin seed components (Lup an 1 or Lup a 1 allergens) to the consumers.

7.3. Management and Prevention of Lupin Allergy

Lupin allergy can be prevented whether the following measures are taken:

1. General awareness of lupin flour as a potential allergen;
2. Food workers and individuals with known allergy should be instructed on the practices of the food industry especially regarding "hidden" allergens;

3. Patients with a known allergy to peanuts should be advised to avoid all products containing lupin until they can be specifically tested;
4. Lupin, when used in a food product, should be included in the list of ingredients in the packaging;
5. The European Union included lupin among allergenic foods for which labeling is mandatory.

A positive skin prick test may not be enough proof for recommendation of stopping the ingestion of a tolerant food. However, having clinical manifestations of allergy then would be mandatory to stop exposure to the allergen. Additionally, administration of an antihistaminic agent might be recommended following medical advice. In the case of anaphylaxis, professional help should be immediately seek, and this may involve standard procedures (i.e., administration of adrenaline either intramuscularly or intravenously, intravenous corticosteroids and antihistamines).

Patients with known lupin flour allergy should be prone to be provided with an emergency kit. In addition, due to the frequency and severity of cross-reactions, patients with peanut allergy should be advised to avoid all products containing lupin.

REFERENCES

Abrams EM, Gerstner TV. (2015). Allergy to cooked, but not raw, peas: a case series and review. *Allergy, Asthma and Clinical Immunology*, 11: 10.

Alvarez-Alvarez J, Guillamón E, Crespo JF, Cuadrado C, Burbano C, Rodríguez J, Fernández C, Muzquiz M. (2005). Effects of extrusion, boiling, autoclaving, and microwave heating on lupin allergenicity. *Journal of Agricultural and Food Chemistry*, 53: 1294–1298.

Arnoldi A, Boschin G, Lammi C, Zanoni C. (2015). The health benefits of sweet lupin seed flours and isolated proteins. *Journal of Functional Foods*, 18: 550–563.

Bähr M, Fechner A, Jahreis G, Kaatz M. (2014). Skin prick test reactivity to lupin in comparison to peanut, pea, and soybean in atopic and non-atopic German subjects: A preliminary cross-sectional study. Immunity, *Inflammation & Disease*, 2: 114–120.

Ballabio C, Penas E, Uberti F, et al. (2013). Characterization of the sensitization profile to lupin in peanut allergic children and assessment of cross-reactivity risk. *Pediatric Allergy and Immunology*, 24: 270–275.

Bansal AS, Bansal RA, Hayman GR, Sanghvi MM. (2014). Variably severe systemic allergic reactions after consuming foods with unlabelled lupin flour: a case series. *Journal of Medical Case Reports,* 8: 55.

Barnett D, Bonham B, Howden ME. (1987). Allergenic cross-reactions among legume foods - an in vitro study. *The Journal of Allergy and Clinical Immunology*, 79: 433−438.

Bernhisel BJ, Sampson HA. (1989). Cross-allergenicity in the legume botanical family in children with food hypersensitivity. *The Journal of Allergy and Clinical Immunology*, 83: 435−440.

Bernhisel BJ, Sampson H, Taylor S. (1989). Cross-allergenicity in the legume botanical family with food hypersensitivity. *The Journal of Allergy and Clinical Immunology*, 84: 701–709.

Brennecke S, Becker WM, Jappe U, Lepp U. (2007). Anaphylactic reaction to lupin flour. *Journal der Deutschen Dermatologischen Gesellschaft* (JDDG), 5: 774–776.

Bublin M, Breiteneder H. (2014). Cross-Reactivity of Peanut Allergens. *Current Allergy and Asthma Reports*, 14: 426.

Campbell CP, Jackson AS, Johnson AR, Thomas PS, Yates DH. (2007). Occupational sensitization to lupin in the workplace: Occupational asthma, rhinitis, and work-aggravated asthma. *The Journal of Allergy and Clinical Immunology*, 119: 1133–1139.

Clark R, Johnson S. (2002). Sensory acceptability of foods with added lupin (*Lupinus angustifolius*) kernel fiber using preset criteria. *Journal of Food Science*, 67: 356−362.

Crespo JF, Pascual C, Burks AW, et al. (1995). Frequency of food allergy in a pediatric population from Spain. *Pediatric Allergy and Immunology,* 6: 39–43.

Crespo JF, Rodriguez J, Vives R, et al. (2001). Occupational IgE-mediated allergy after exposure to lupin seed flour. *Journal of Allergy and Clinical Immunology,* 108: 295–297.

Crespo J, Rodriguez J. (2004). Diagnosis and therapy of food allergy. *Molecular Nutrition & Food Research*, 48: 347–355.

Chruszcz M, Maleki SJ, Majorek KA, Demas M, Bublin M, Solberg R, Hurlburt BK, Ruan S, Mattison CP, Breiteneder H, Minor W. (2011). Structural and Immunologic Characterization of Ara h 1, a Major Peanut Allergen. *The Journal of Biological Chemistry*, 286: 39318–39327.

de Jong NW, van Maaren MS, Vlieg-Boersta BJ, Dubois AE, de Groot H, Gerth van Wijk R. (2010). Sensitization to lupin flour: is it clinically relevant? *Clinical and Experimental Allergy*, 40: 1571–1577.

Dooper MM, Holden L, Faeste CK, Thompson KM, Egaas E. (2006). Monoclonal antibodies against the candidate lupin allergens alpha-conglutin and beta-conglutin. *International Archives of Allergy and Immunology,* 143: 49–58.

Duranti M. (2006). Grain legume proteins and nutraceutical properties. *Fitoterapia,* 77: 67–82.

Duranti M, Consonni A, Magni C, Sessa F, Scarafoni A. (2008). The major proteins of lupin seed: Characterization and molecular properties for use as functional and nutraceutical ingredients. *Trends in Food Science & Technology*, 19: 624–633.

Egorov TA, Odintsova TI, Musolyamov AK, Fido R, Tatham AS, Shewry PR. (1996). Disulphide structure of a sunflower seed albumin: conserved and variant disulphide bonds in the cereal prolamin superfamily. *FEBS Letters*, 396: 285–8.

Eigenmann PA, Burks AW, Bannon GA, Sampson HA. (1996). Identification of unique peanut and soy allergens in sera adsorbed with cross-reacting antibodies. *Journal of Allergy and Clinical Immunology,* 98: 969–78.

Faeste CK, Lovik M, Wiker HG, Egaas E. (2004). A case of peanut cross-allergy to flour in hot dog bread. *International Archives of Allergy and Immunology,* 135: 36-39.

Fæste, C. K. (2009) Lupin Allergen Detection, in Molecular Biological and Immunological Techniques and Applications for Food Chemists (eds B. Popping, C. Diaz-Amigo and K. Hoenicke), John Wiley & Sons, Inc., Hoboken, NJ, USA. doi: 10.1002/9780470637685.ch22.

Fiocchi A, Sarratud P, Terracciano L, Vacca E, Bernardini R, Fuggetta D, Ballabio C, Duranti M, Magni C, Restani P. (2009). Assessment of the tolerance to lupin-enriched pasta in peanut-allergic children. *Clinical and Experimental Allergy*, 39: 1045-1051.

Gayraud J, Mairesse M, Fontaine JF, Thillay A, Leduc V, Rancé F, Parisot L, Moneret-Vautrin DA. (2009). The prevalence of sensitization to lupin flour in France and Belgium: a prospective study in 5,366 patients, by the Allergy Vigilance Network. *European Annals of Allergy and Clinical Immunology*, 41: 17–22.

Goggin DE, Mir G, Smith WB, Stuckey M, Smith PMC. (2008). Proteomic analysis of lupin seed proteins to identify conglutin b as an allergen, Lup an 1. *Journal of Agricultural and Food Chemistry*, 56: 6370–6377.

Guarneri F, Guarneri C, Benvenga S. (2005). Identification of potentially cross-reactive peanut-lupin proteins by computer-assisted search for amino acid sequence homology. *International Archives of Allergy and Immunology*, 138: 273–7.

Guillamón E, Burbano C, Cuadrado C, Muzquiz M, Pedrosa M. M, Sánchez M, Cabanillas B, Crespo J. F, Rodriguez J, Haddad J, Allaf K. (2008). Effect of an instantaneous controlled pressure drop on in vitro allergenicity to lupins (Lupinus albus var Multolupa). *International Archives of Allergy and Immunology*, 145: 9–14.

Guillamón, E, Rodríguez, J, Burbano, C, Muzquiz, M, Pedrosa, MM, Cabanillas, B, Crespo, JF, Sancho, AI, Mills, EN, Cuadrado, C. (2010). Characterization of lupin major allergens (*Lupinus albus* L.). *Molecular Nutrition and Food Research, 54*(11): 1668-1676.

Hefle SL, Lemanske RF Jr, Bush RK. (1994). Adverse reaction to lupin fortified pasta. *Journal of Allergy and Clinical Immunology*, 94: 167–72.

Holden L, Sletten GBG, Lindvik H, Faeste CK, Dooper MMBW. (2008). Characterization of IgE binding to lupin, peanut and almond with sera from lupin-allergic patients. *International Archives of Allergy and Immunology*, 146: 267–76.

Islam S, Yan G, Appels R, et al. (2012). Comparative proteome analysis of seed storage and allergenic proteins among four narrow-leafed lupin cultivars. *Food Chemistry*, 135: 1230-1238.

Jappe U, Brennecke S, Lepp U, Becker W. (2007). Lupin allergy: monovalent sensitisation to lupin seed flour without cross-reactivity to legumes. *Allergy*, 62: 368.

Lallès JP, Peltre G. (1996). Biochemical features of grain legume allergens in humans and animals. *Nutrition Reviews*, 54: 101–107.

Leduc V, Moneret-Vautrin DA, Guérin L. (2002). Allergenicity of lupin flour. *Allergie et Immunologie*, 34: 213-217.

Lima-Cabello, E, Robles-Bolivar, P, Alche, JD, Jimenez-Lopez, JC. (2016). Narrow-leafed lupin β-conglutin proteins epitopes identification and molecular features analysis involved in cross-allergenicity to peanut and other legumes. *Genomics and Computational Biology,* 2(1): e29, doi: 10.18547/gcb.2016.vol2.iss1.e29.

Lima-Cabello, E, Alche, V, Foley, RC, Andrikopoulos, S, Morahan, G, Singh, KB, Alche, JD, Jimenez-Lopez, JC. (2016). Narrow-leafed lupin (*Lupinus angustifolius* L.) β-conglutin proteins modulate the insulin signalling pathway as potential type 2 diabetes treatment and inflammatory-related disease amelioration. *Molecular Nutrition and Food Research*, 61(5), doi: 10.1002/mnfr.201600819.

Lindvik H, Holden L, Lovik M, Cvancarova M, Halvorsen R. (2008). Lupin sensitization and clinical allergy in food allergic children in Norway. *Acta Paediatrica*, 97: 91-95.

Lucas MM, Stoddard FL, Annicchiarico P, Frías J, Martínez-Villaluenga C, Sussmann D, Duranti M, Seger A, Zander PM, Pueyo JJ. (2015).

The future of lupin as a protein crop in Europe. *Frontiers in Plant Science,* 6: 705.

Magni C, Herndl A, Sironi E et al. (2005). One- and two-dimensional electrophoretic identification of IgE-binding polypeptides of Lupinus albus and other legume seeds. *Journal of Agricultural and Food Chemistry,* 53: 4567–4571.

Magni C, Herndl A, Sironi E, Scarafoni A. et al. (2005). One and two-dimensional electrophoretic identification of IgE-binding polypeptides of Lupinus albus and other legume seeds. *Journal of Agricultural and Food Chemistry*, 53: 4567–4571.

Magni C, Scarafoni A, Herndl A, Sessa F, Prinsi B, Espen L, Duranti M. (2007). Combined 2D electrophoretic approaches for the study of white lupin mature seed storage proteome. *Phytochemistry*, 68: 997-1007.

Magni C, Sessa F, Tedeschi G, Negri A, Scarafoni A, Consonni A, Duranti M. (2012). Identification in lupin seed of a serine-endopeptidase activity cleaving between twin arginine pairs and causing limited proteolysis of seed storage proteins. *Molecular Plant*. 5: 1011-1019.

Matheu V, de Barrio M, Sierra Z, Gracia-Bara MT, Tornero P, Baeza ML. (1999). Lupin-induced anaphylaxis. *Annals of Allergy, Asthma & Immunology,* 83: 406-408.

Mazeyrat R., Thibault M., Asensi D., Ponvert C.: Sensitisation to lupin flour: Evaluation of risk in peanut-allergic children. In: Abstract Book of the 23rd European Academy of Allergology and Clinical Immunology, Amsterdam, ISBN 3-980961-3-7 300.

Moneret-Vautrin DA, Guérin L, Kanny G, Flabbee J, Frémont S, Morisset M. (1999). Cross-allergenicity of peanut and lupin: the risk of lupin allergy in patients allergic to peanuts. *Journal of Allergy and Clinical Immunology,* 104: 883–888.

Moreno-Ancillo A, Gil-Adrados AC, Dominguez-Noche C, Cosmes PM. (2005). Lupin inhalation induced asthma in a child. *Pediatric Allergy and Immunology,* 16: 542–544.

Nicolaou N, Custovic A. Molecular diagnosis of peanut and legume allergy. (2011). *Current Opinion in Allergy & Clinical Immunology,* 11: 222–228.

Novembre E, Moriondo M, Bernardini R, Azzari C, Rossi ME, Vierucci A. (1999). Lupin allergy in a child. *Journal of Allergy and Clinical Immunology,* 103: 1214–1216.

Parisot L, Aparicio C, Moneret-Vautrin DA, Guerin L. (2001). Allergy to lupin flour. *Allergy,* 56: 918–919.

Peeters KA, Koppelman SJ, Penninks AH, Lebens A, Bruijnzeel-Koomen CA, Hefle SL, Taylor SL, van Hoffen E, Knulst AC. (2009). Clinical relevance of sensitization to lupin in peanut sensitized adults. *Allergy,* 64: 549–555.

Peeters KA, Nordlee JA, Penninks AH, Chen L, Goodman RE, Bruijnzeel-Koomen CA, Hefle SL, Taylor SL, Knulst AC. (2007). Lupin allergy: not simply cross-reactivity with peanut or soy. *Journal of Allergy and Clinical Immunology*, 120: 647-53.

Pereira MJ, Belver MT, Pascual CY, Martin Esteban M. (2002). The allergenic significance of legumes. *Allergologia et Immunopathologia,* 30: 346–353.

Poltronieri P, Cappello MS, Dohmae N, Conti A, Fortunato D, Pastorello EA, Ortolani C, Zacheo G. (2002). Identification and characterisation of the IgE-binding proteins 2S albumin and conglutin gamma in almond (Prunus dulcis) seeds. *International Archives of Allergy and Immunology*, 128: 97-104.

Pope AM, Patterson R, Burge H, & editors. (1993). Indoor Allergens: Assessing and Controlling Adverse Health Effects. Magnitude and Dimensions of Sensitization and Disease Caused by Indoor Allergens. Institute of Medicine (US) Committee on the Health Effects of Indoor Allergens. Washington (DC): National Academies Press (US). ISBN-10: 0-309-04831-1.

Prieto A, Razzak E, Lindo DP, Alvarez-Perea A, Rueda M, Baeza ML. (2010). Recurrent anaphylaxis due to lupin flour: primary sensitization through inhalation. *Journal of Investigational Allergology and Clinical Immunology,* 20: 76-79.

Quaresma RR, Viseu R, Martins LM, Tomaz E, Inácio F. (2007). Allergic primary sensitization to lupin seed. *Allergy,* 62: 1473-1474.

Radcliffe M, Scadding G, Brown HM. (2005). Lupin flour anaphylaxis. *The Lancet*, 365:1360.

Reis AM, Fernandes NP, Marques SL, Paes MJ, Sousa S, Carvalho F, Conde T, Trindade M. (2007). Lupin sensitisation in a population of 1,160 subjects. *Allergologia et Immunopathologia*, 35: 162-163.

Remington B, Baumert J, Bijlsma S, Houben G, Kruizinga A, Taylor S, Westerhout J. (2016). Food allergy population thresholds: 2011 - 2016 update of international DBPCFC database for food allergen risk assessment. 4th food allergy and anaphylaxis meeting 2016. *European Academy of Allergy & Clinical Immunology* (EAACI), Rome.

Rojas-Hijazo B, Garces MM, Caballero ML, Alloza R, Moneo I. (2006). Unsuspected lupin allergens hidden in food. *International Archives of Allergy and Immunology*, 141: 47–50.

Rosales MJ, Marco FM, Raga E, Sempere JM, Diaz P. (2002). Lupin seed anaphylaxis. A study of lupin allergens. [Poster: XXI Congress of EAACI] *Allergy,* 57 Suppl 73:85-105.

Rotiroti G, Skypala I, Senna G, Passalacqua G. (2007). Anaphylaxis due to lupin flour in a celiac patient. *Journal of Investigational Allergology and Clinical Immunology*, 17: 204–205.

Sanz ML, de Las Marinas MD, Fernandez, J, et al. (2010). Lupin allergy: a hidden killer in the home. *Clinical & Experimental Allergy*, 40: 1461-1466.

Scarafoni A, Ronchi A, Duranti M. (2009). A realtime PCR method for the detection and quantification of lupin flour in wheat flour-based matrices. *Food Chemistry*, 115: 1088–1093.

Shaw J, Roberts G, Grimshaw K, White S, Hourihane J. (2008). Lupin allergy in peanut-allergic children and teenagers. *Allergy,* 63: 370-373.

Smith WB, Gillis D, Kette FE. (2004). Lupin: a new hidden food allergen. *The Medical Journal of Australia*, 181: 219-220.

Terruzzi I, Senesi P, Magni C, Montesano A, Scarafoni A, Luzi L, & Duranti M. (2011). Insulin-mimetic action of conglutin-gamma, a lupin seed protein, in mouse myoblasts. *Nutrition, Metabolism, and Cardiovascular Diseases*, 21: 197–205.

Verma AK, Kumar S, Das M, et al. (2013). A Comprehensive Review of Legume Allergy. *Clinical Reviews in Allergy & Immunology*, 45: 30–46.

Verma AK, Kumar S, Das M, Dwivedi PD. (2012). Impact of thermal processing on legume allergens. *Plant Foods for Human Nutrition*, 67: 430-441.

Wait R, Gianazza E, Brambilla D, Eberini I, Morandi S, Arnoldi A, Sirtori CR. (2005). Analysis of Lupinus albus storage proteins by two-dimensional electrophoresis and mass spectrometry. *Journal of Agricultural and Food Chemistry,* 53: 4599-4606.

Wassenberg J, Hofer M. (2007). Lupin-induced anaphylaxis in a child without known food allergy. *Annals of Allergy, Asthma & Immunology*, 98: 589-590.

Wüthrich B. (2008) Anaphylactic reaction to lupin flour because of a primary sensitization. *Allergy,* 63: 476-477.

Yman L. (1982). Botanical relations and immuno-logical cross-reactions in pollen allergy. 2nd edition. *Pharmacia Diagnostics* AB. Uppsala, Sweden. ISBN 91-970475-09.

About the Editors

Dr. Alfonso Clemente
Tenured Scientist CSIC
Department of Physiology and Biochemistry of the Animal Nutrition;
Estacion Experimetal del Zaidin;
Spanish National Research Council (CSIC);
Granada, Spain

Dr. Alfonso Clemente is a staff scientist of the Spanish National Research Council, working at the Estación Experimental del Zaidín (EEZ-CSIC, Granada, Spain). He has been working in legume seeds for the last twenty years, being involved since then in several national and international related projects. Alfonso Clemente joined different labs (Institute of Food Research, 1999-200; John Innes Centre, 2000-2002; Sainsbury Laboratory, 2003-2004) in UK to broaden his laboratory skills and scientific knowledge. Alfonso Clemente is currently President of the Spanish Legume Association (AEL, www.leguminosas.es) having strong interaction with a relevant network of scientists at national and international level as well as agricultural associations and agri-food companies in the field. As President of AEL, he is leading a number of initiatives in the International Year of Legumes (IYP2016) in collaboration with CSIC, INIA, Triptolemos Foundation, Mensa Civica, and other relevant organizations. In addition, and focused on gastrointestinal health, Dr. Clemente is Board member of the Spanish Society of Probiotics and Prebiotics (www.sepyp.es). He is author of more than 120 scientific manuscripts and editorial board member of World Journal of Gastroenterology and Frontiers in Bioscience, among others.

Dr. Jose C. Jimenez-Lopez
Senior Research Fellow CSIC
Department of Biochemistry, Cell and Molecular Biology of Plants;
Estacion Experimetal del Zaidin;
Spanish National Research Council (CSIC);
Granada, Spain

Dr. Jose C. Jimenez-Lopez graduated by the University of Granada (UGR), Spain in Biochemistry and Molecular Biology (1998), and Biological Sciences (2001), and he completed a Master degree in

Agricultural Sciences (2004) at UGR and The Spanish National Research Council (CSIC). Jose was awarded a PhD degree in 2008 by CSIC. Soon after, he moved to Purdue University (USA) to develop his post-doctoral research in Plant Biology until 2011. In 2012 he was awarded an European Grant-Project from the Marie Curie program (FP7-PEOPLE-2011-IOF) to work at The University of Western Australia (2012'2014) in nutraceutical aspects and health benefits of the legume *lupinus angustifolius*, and at CSIC (2014-2015) in lupin molecular allergy. Currently he holds a Senior Research Fellow position at CSIC thanks to a grant-project from the Spanish Government (MINECO) – Ramon y Cajal research program, working in the functional characterization of proteins from plant reproductive (pollen and seed) tissues in plant species of agro-industrial interest as legumes and olive. Dr. Jimenez-Lopez has co-authored more than 70 scientific manuscripts including articles in peer-reviewed journals and book chapters. He has presented his research accomplishments in more than 65 international conferences, and currently is an active member of seven international scientific societies as International Legume Society, serving as a referee for different reputed peer-reviewed journals, and also as a scientific editor for several book editorials.

INDEX

C

D

E

M

N

O

P

R

S

T

V

W

Z